Applied Physics for Electronic Technology

Applied Physics for Electronic Technology

A problem solving approach

ANDREW A LEVEN BSc (Hons), MSc, CEng, MIEE, MIP

A member of the Hodder Headline Group
LONDON • SYDNEY • AUCKLAND

To my wife Lorna
and the memory of Robert Kirk

First published in Great Britain 1998 by
Arnold, a member of the Hodder Headline Group,
338 Euston Road, London NW1 3BH
http://www.arnoldpublishers.com

Whilst the advice and information in this book are believed to be true and
accurate at the date of going to press, neither the author nor the publisher
can accept any legal responsibility or liability for any errors or omissions
that may be made.

British Library Cataloguing in Publication Data
A catalogue record for this book is available from the British Library

Library of Congress Cataloging-in-Publication
A catalogue record for this book is available from the Library of Congress

ISBN 0 340 69155 7

Commissioning Editor: Sian Jones
Production Editor: James Rabson
Production Controller: Priya Gohil
Cover design: Terry Griffiths

Typeset in 10/12 pt Times by Academic & Technical Typesetting, Bristol
Printed and bound in Great Britain by J W Arrowsmith, Bristol

Contents

1

DC networks

1.1 Introduction

Many applications in electronic technology require the implementation of d.c. networks. This may involve the use of Ohm's law in the simplest resistance d.c. circuits to the more complex circuits requiring other theorems. The circuit in Fig. 1.1 illustrates an attenuation problem between a source cable and a load cable. In order to attenuate a current I, a resistive network is inserted between the cables. Such a network is easily solved by the use of simple circuit analysis. Such circuits will be examined in this chapter.

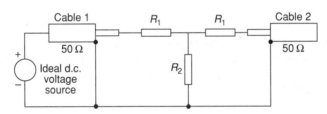

Figure 1.1

1.2 Electrical properties

All matter consists of atoms and the model used to understand the structure of the atom was called the Rutherford atom. Modern physics now considers the energy level model, but this will be discussed in Chapter 3. The Rutherford model consists of a central core or nucleus which contains charged particles called protons and neutral particles called neutrons. Orbiting about this nucleus at different distances are much smaller negatively charged particles called electrons. An atom normally contains equal numbers of protons and electrons. Under these conditions the atom is neutral, i.e. it has no charge.

As the electrons in the outer orbits are more loosely bound by the electric field between protons and electrons, some of these can be removed from the atom and travel to another atom. The neutral atoms are no longer neutral but become ions, i.e. a particle which has more or less electrons than its neutral state. This is shown in Fig. 1.2 in which for a moment atom A is positively charged and atom B is negatively charged.

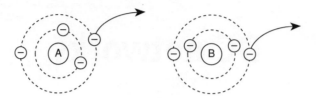

Figure 1.2

A simple example of this occurs when certain materials are rubbed, e.g. polythene and acetate. Generally, however, an electromotive force such as a battery or generator is required to generate large electron flows which is referred to as electricity. Associated with electricity are five basic properties which are required to be understood in order to solve d.c. networks:
1. charge
2. current
3. potential difference
4. resistance
5. energy and power.

Charge

The quantity of electrical charge is given the symbol Q and is measured in a unit called the coulomb (C). In any process in which charge is transferred from one body to another, the total charge is conserved. Mathematically the charge is related to current and time as

$$Q = It \tag{1.1}$$

Current (*I*)

An electric current is a flow of charged particles. In metals the charge is carried by electrons while in semiconductors ions and hole flow are involved. The unit of current is the ampere and its direction by convention is taken from points of higher potential to points of lower potential, i.e. less positive. However, when the current is being carried by electrons the electron flow is in the opposite direction.

Potential difference

When a mass is raised through a height, work is done on the mass and is equal to the potential energy gained by the mass. In a similar way when an electron is placed in an electric field as shown in Fig. 1.3, work is done in moving the electron from A to B and hence the electron gains energy. Note that in

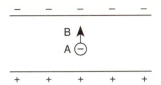

Figure 1.3

Fig. 1.3 the electron has a greater amount of energy at B and that a difference in potential exists between A and B. This is generally referred to as potential difference. There is an important point to notice here concerning the difference between potential and p.d. Generally potential is referred to as positive or negative, but by convention potentials below earth (which is a neutral reference point) are negative while potentials above earth are positive. Figure 1.4 illustrates different potentials which are measured in volts (V) with respect to earth. However, the p.d. between levels in each case is respectively 150, 60 and 0 V.

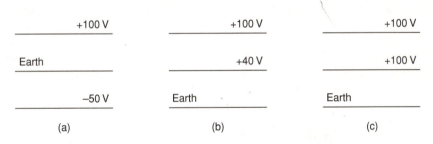

Figure 1.4

Resistance

When two points having a potential difference between them are joined by a conductor, the ease with which the charge flows between the points depends on the nature of the conducting material. Metals such as copper, aluminium and silver are good conductors and have low resistance to flow while materials such as glass have high resistances. Electrical resistance is measured in ohms (Ω).

Energy

Electrical energy is given as

$$\text{Energy} = VIt \text{ joules (J)} \tag{1.2}$$

while power is the rate at which work is done or energy is transferred:

$$\text{Power} = \frac{\text{Energy}}{\text{Time}} = \frac{E}{t}$$

$$P = \frac{VIt}{t} = VI \text{ watts (W)} \tag{1.3}$$

Ohm's law

The electrical resistance of a conductor is a measure of the p.d. required to maintain a current through it. A relationship exists between these three properties which is known as Ohm's law:

$$R = \frac{V}{I} \qquad (1.4)$$

This law enables two further power equations to be derived from equation (1.3). From $V = IR$, also $P = VI$

$$\therefore \; P = I^2 R \qquad (1.5)$$

Also $I = V/R$

$$\therefore \; P = \frac{V^2}{R} \qquad (1.6)$$

1.3 Resistor combinations

The majority of d.c. electrical circuits have a large number of resistors connected in many different configurations. The three basic ways of connecting resistors are (1) series, (2) parallel and (3) series/parallel arrangements. All circuits can be reduced to these configurations.

1.4 Series circuits

In the type of circuit shown in Fig. 1.5 the following points should be noted:

- The current through each resistor is the same.
- The sum of the individual p.d.s is equal to the supply voltage:

$$V = V_1 + V_2 + V_3$$

- The total resistance in the circuit is equal to the sum of the individual resistances:

$$R_T = R_1 + R_2 + R_3$$

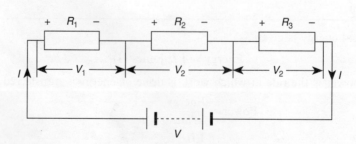

Figure 1.5

Example 1.1

In the circuit in Fig. 1.6 determine the following:

(a) the value of I
(b) the p.d. across each resistor
(c) the power taken by each resistor.

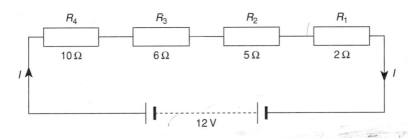

Figure 1.6

Solution

(a) $$R_T = R_1 + R_2 + R_3 + R_4 = 2 + 5 + 6 + 10 = 23\,\Omega$$

$$\therefore\ I = \frac{V}{R} = \frac{12}{23} = 0.5217$$

(b) Since the current is the same through each resistor then

$$V_1 = IR_1 = 0.5217 \times 2 = 1.04\,\text{V}$$
$$V_2 = IR_2 = 0.5217 \times 5 = 2.61\,\text{V}$$
$$V_3 = IR_3 = 0.5217 \times 6 = 3.13\,\text{V}$$
$$V_4 = IR_4 = 0.5217 \times 10 = 5.22\,\text{V}$$

Note:

(c) $$P_1 = I^2 R_1 = (0.5217)^2 \times 2 = 0.54\,\text{W}$$
$$P_2 = I^2 R_2 = (0.5217)^2 \times 5 = 1.36\,\text{W}$$
$$P_3 = I^2 R_3 = (0.5217)^2 \times 6 = 1.63\,\text{W}$$
$$P_4 = I^2 R_4 = (0.5217)^2 \times 10 = 2.72\,\text{W}$$

Example 1.2

Two heater elements are connected as shown in Fig. 1.7. Calculate the value of the series resistor R_s which will enable the two heaters to function with their correct ratings.

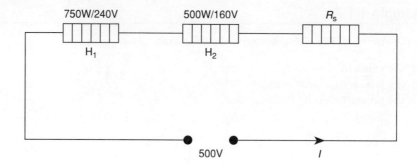

Figure 1.7

Solution
The resistance of H_1 is

$$R_{H_1} = \frac{V_{H_1}^2}{P_{H_1}} = \frac{(240)^2}{750} = 76.8 \,\Omega$$

$$R_{H_2} = \frac{V_{H_2}^2}{P_{H_2}} = \frac{(160)^2}{500} = 51.2 \,\Omega$$

∴ The total resistance of the heating elements is $128 \,\Omega$.
The rated current of H_1 is

$$I = \frac{P_1}{V_1} = \frac{750}{240} = 3.125 \,\text{A}$$

The rated current of H_2 is

$$I = \frac{P_2}{V_2} = \frac{500}{160} = 3.125 \,\text{A}$$

The total resistance required in the circuit must be

$$R_T = \frac{V}{I} = \frac{500}{3.125} = 160 \,\Omega$$
$$R_s = R_T - 128 \,\Omega = 160 - 128 = 32 \,\Omega$$

1.5 Parallel circuits

Parallel circuits can take many configurations and it is not always obvious when components are actually in parallel. Two simple rules enable this to be determined.

In Fig. 1.8 the following can be stated:

1. By conservation of charge

$$I = I_1 + I_2 + I_3$$

i.e. the supply current equals the sum of the individual currents.

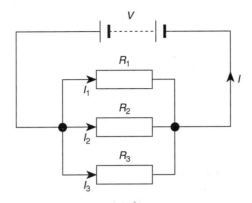

Figure 1.8

2. The p.d. across conductors connected in parallel is the same.

These statements enable the effective resistance (R_E) to be calculated as follows:

$$\frac{V}{R_E} = \frac{V}{R_1} + \frac{V}{R_2} + \frac{V}{R_3}$$

$$\frac{1}{R_E} = \frac{1}{R_1} + \frac{1}{R_2} + \frac{1}{R_3}$$

(1.7)

For the special case of two conductors in parallel we have

$$\frac{1}{R_E} = \frac{1}{R_1} + \frac{1}{R_2} = \frac{R_1 + R_2}{R_1 R_2}$$

Taking the reciprocal gives

$$R_E = \frac{R_1 R_2}{R_1 + R_2}$$

(1.8)

Note that the effective resistance of any number of conductors or resistors connected in parallel is always less than the smallest resistor in the parallel network.

Example 1.3

From Fig. 1.9, determine the following:

(a) the supply current (I)
(b) the current in each branch
(c) the power rating of each resistor.

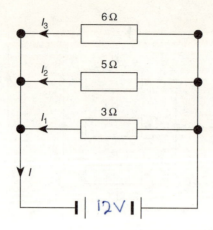

Figure 1.9

Solution

(a)

$$\frac{1}{R_E} = \frac{1}{R_1} + \frac{1}{R_2} + \frac{1}{R_3}$$

$$= \frac{1}{3} + \frac{1}{5} + \frac{1}{6}$$

$$= \frac{10 + 6 + 5}{30} = \frac{21}{30}$$

$$\therefore \ R_E = \frac{30}{21} = 1.429 \, \Omega$$

$$I = \frac{V}{R_E} = \frac{12}{1.429} = 8.397 \, \text{A}$$

(b)

$$I_1 = \frac{V}{R_1} = \frac{12}{3} = 4 \, \text{A}$$

$$I_2 = \frac{V}{R_2} = \frac{12}{5} = 2.4 \, \text{A}$$

$$I_3 = \frac{V}{R_3} = \frac{12}{6} = 2 \, \text{A}$$

(c)

$$P_1 = I_1^2 R_1 = (4)^2 \times 3 = 48 \, \text{W}$$

$$P_2 = I_2^2 R_2 = (2.4)^2 \times 5 = 28.8 \, \text{W}$$

$$P_3 = I_3^2 R_3 = (2)^2 \times 6 = 24 \, \text{W}$$

Hence resistors with wattage ratings greater than these values would be appropriate. The currents are calculated as

$$I_1 = \tfrac{1}{4} \times 4.8 = 1.2 \, \text{A}$$

$$I_2 = \tfrac{5}{3} \times 4.8 = 3 \, \text{A}$$

$$I_3 = \tfrac{1}{8} \times 4.8 = 0.6 \, \text{A}$$

Example 1.4

A d.c. generator has to provide 3 A through a 2 Ω resistor on an open circuit as shown in Fig. 1.10. Determine:

(a) the value of the potentiometer (RV_1) under these conditions;
(b) the value of the current through branch 2 when a resistive load of 5 Ω is connected across the 2 Ω resistor and RV_1 is set as in (a);
(c) the total current required from the generator in both cases.

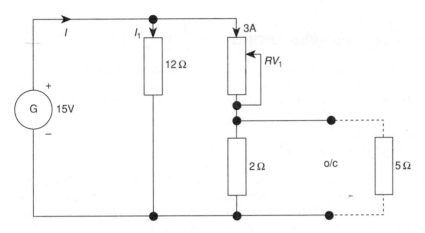

Figure 1.10

Solution

(a)
$$I = \frac{V}{RV_1 + 2}$$

$$RV_1 = \frac{V - 2I}{I}$$

$$= \frac{15 - (2 \times 3)}{3}$$

$$= \frac{15 - 6}{3} = 3\,\Omega$$

(b) The 2 Ω and 5 Ω resistors are in parallel, hence, from equation (1.8)

$$R_E = \frac{R_1 \times R_2}{R_1 + R_2}$$

$$= \frac{5 \times 2}{5 + 2} = 1.428\,\Omega$$

The circuit for branch 2 now becomes a series resistance of $3\,\Omega + 1.428\,\Omega = 4.428\,\Omega$

$$\therefore\ I_2 = \frac{15}{4.428} = 3.387\,\text{A}$$

(c)
$$I_1 = \frac{15}{12} = 1.25\,\text{A}$$

Without the load,
$$I = I_1 + I_2 = 1.25 + 3 = 4.25\,\text{A}$$

With the load,
$$I = 1.25\,\text{A} + 3.387 = 4.637\,\text{A}$$

1.6 Series/parallel circuits

These circuits are by far the most frequently used in industrial applications, but before tackling them the techniques of current and voltage division will be explained as these are both used in series/parallel circuits.

1.7 Voltage division

It is often required to take a portion of the total voltage in a circuit and this can be done by using either a potentiometer or fixed resistors. If a load is connected across the output of the potential divider the output voltage changes as the circuit is now a series/parallel arrangement (Fig. 1.11).

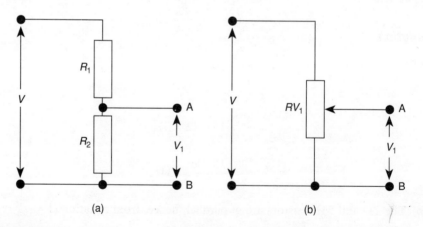

(a)   (b)

Figure 1.11

1.8 Current division

As has been seen in the parallel circuit, whenever the current in any circuit encounters a branch, it splits and the individual current values can be calculated as in Example 1.3. However, a simple current division rule can be applied and this is best shown by an example.

Example 1.5

It is required to find all the currents indicated on Fig. 1.12.

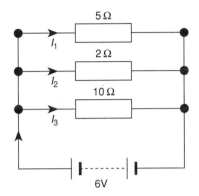

Figure 1.12

Solution

In order to find the total current the equivalent resistance (R_E) has to be found:

$$\frac{1}{R_E} = \frac{1}{R_1} + \frac{1}{R_2} + \frac{1}{R_3}$$

$$= \frac{1}{5} + \frac{1}{2} + \frac{1}{10}$$

$$= \frac{2+5+1}{10} = \frac{8}{10} \qquad (1.9)$$

$$R_E = \frac{10}{8} = 1.25\,\Omega$$

Hence

$$I = \frac{V}{R_E} = \frac{6}{1.25} = 4.8\,\text{A}$$

In order to find how this current splits we consider expression (1.9). The numerator tells us that the total current of 4.8 A splits into eight portions with two portions going through the 5 Ω resistor, five portions through the 2 Ω resistor and one portion going through the 10 Ω resistor.

The currents are calculated as

$$I_1 = \frac{1}{4} \times 4.8 = 1.2\,\text{A}$$

$$I_2 = \frac{5}{8} \times 4.8 = 3\,\text{A}$$

$$I_3 = \frac{1}{8} \times 4.8 = 0.6\,\text{A}$$

These principles will now be investigated in series/parallel arrangements.

Example 1.6

From Fig. 1.13(a), determine:

(a) all the currents
(b) the voltages across each resistor.

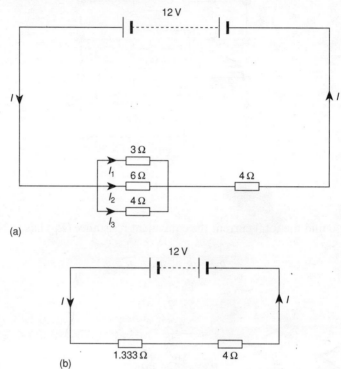

(a)

(b)

Figure 1.13

Solution

(a)

$$\frac{1}{R_E} = \frac{1}{3} + \frac{1}{6} + \frac{1}{4}$$

$$= \frac{4 + 2 + 3}{12} = \frac{9}{12}$$

$$\therefore \ R_E = 1.333 \, \Omega$$

$$I = \frac{12}{5.333} = 2.25 \, \text{A}$$

By current division

$$I_1 = \frac{4}{9} \times 2.25 = 1 \, \text{A}$$

$$I_2 = \frac{2}{9} \times 2.25 = 0.5 \, \text{A}$$

$$I_3 = \frac{3}{9} \times 2.25 = 0.75 \, \text{A}$$

(b) The voltage across each resistor in the parallel branch is

$$V = I_1 R_1 = I_2 R_2 = I_3 R_3$$
$$\therefore\; V = 1 \times 3 = 3\,\text{V}$$

By voltage division the voltage across the $4\,\Omega$ resistor is

$$12 - 3 = 9\,\text{V}$$

Using Fig. 1.13(b) gives a similar answer

$$\left(\frac{4}{4 + 1.333}\right) \times 12 = 9\,\text{V}$$

Example 1.7

From Fig. 1.14, determine:

(a) all the currents indicated
(b) the voltage across $3\,\Omega$ and $6\,\Omega$ resistors
(c) the power rating of the $4\,\Omega$ resistor.

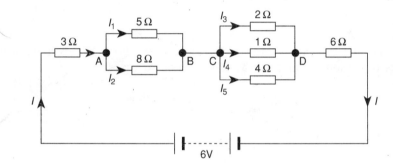

Figure 1.14

Solution

(a) The resistance between A and B is

$$R_{\text{AB}} = \frac{5 \times 8}{5 + 8} = 3.076\,\Omega$$

The resistance between C and D is

$$\frac{1}{2} + \frac{1}{1} + \frac{1}{4} = \frac{2 + 4 + 1}{4}$$
$$\therefore\; R_{\text{CD}} = 0.571\,\Omega$$

The equivalent circuit is shown in Fig. 1.15.

$$\therefore\; I = \frac{6}{3 + 3.076 + 0.571 + 6}$$
$$= 0.474\,\text{A}$$

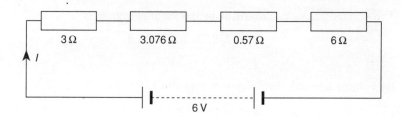

Figure 1.15

For two resistors in parallel

$$\frac{1}{R_E} = \frac{1}{R_1} + \frac{1}{R_2} = \frac{R_2 + R_1}{R_1 R_2}$$

Therefore, currents through R_1 and R_2 are

$$I_1 = \left(\frac{R_2}{R_1 + R_2}\right) \times I$$

$$I_2 = \left(\frac{R_1}{R_1 + R_2}\right) \times I$$

Note: Divide the value of the resistor through which you do not wish to know the current by the sum of the two resistors. This is a frequently used relationship and will be used in later examples.

$$\therefore I_1 = \left(\frac{R_2}{R_1 + R_2}\right) \times I$$

$$= \left(\frac{8}{8 + 5}\right) \times 0.474 = 0.292 \, \text{A}$$

$$I_2 = \left(\frac{R_1}{R_1 + R_2}\right) \times I$$

$$= \left(\frac{5}{2 + 5}\right) \times 0.474 = 0.182 \, \text{A}$$

$$I_3 = \frac{2}{7} \times 0.474 = 0.135 \, \text{A}$$

$$I_4 = \frac{4}{7} \times 0.474 = 0.271 \, \text{A}$$

$$I_5 = \frac{1}{7} \times 0.474 = 0.068 \, \text{A}$$

(b) Voltage across $3 \, \Omega$ resistor is

$$V = 0.474 \times 3 = 1.422 \, \text{V}$$

Voltage across $6 \, \Omega$ resistor is

$$V = 0.474 \times 6 = 2.844 \, \text{V}$$

(c)
$$P = I^2 R = (0.068)^2 \times 4$$
$$= 0.0185 \, \text{W} = 18.5 \, \text{mW}$$

Example 1.8

Determine the following in Fig. 1.16:

(a) all the currents
(b) the voltage across all the resistors
(c) suitable power ratings for each resistor.

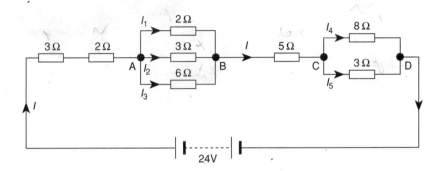

Figure 1.16

Solution

(a)
$$\frac{1}{R_{AB}} = \frac{1}{2} + \frac{1}{3} + \frac{1}{6}$$
$$= \frac{3 + 2 + 1}{6}$$
$$\therefore \ R_{AB} = 1 \, \Omega$$
$$R_{CD} = \frac{8 \times 3}{8 + 3} = \frac{24}{11} = 2.182 \, \Omega$$

The equivalent circuit is shown in Fig. 1.17.

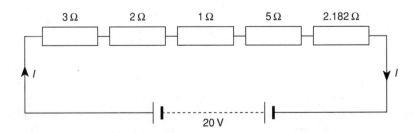

Figure 1.17

$$\therefore \; R_E = 3 + 2 + 1 + 5 + 2.182 = 13.182 \, \Omega$$

$$\therefore \; I = \frac{20}{13.182} = 1.517 \, A$$

$$I_1 = \frac{3}{6} \times 1.517 = 0.759 \, A$$

$$I_2 = \frac{2}{6} \times 1.517 = 0.506 \, A$$

$$I_3 = \frac{1}{6} \times 1.517 = 0.252 \, A$$

$$I_4 = \frac{3}{8+3} \times 1.517 = 0.474 \, A$$

$$I_5 = \frac{8}{8+3} \times 1.517 = 1.103 \, A$$

(b)

$3 \, \Omega$ resistor: $1.517 \times 3 = 4.55 \, V$

$2 \, \Omega$ resistor: $1.517 \times 2 = 3.03 \, V$

$6 \, \Omega$ resistor: $0.252 \times 6 = 1.51 \, V$

$3 \, \Omega$ resistor: $0.506 \times 3 = 1.518 \, V$

$2 \, \Omega$ resistor: $0.759 \times 2 = 1.518 \, V$

$5 \, \Omega$ resistor: $1.517 \times 5 = 7.59 \, V$

$8 \, \Omega$ resistor: $0.414 \times 8 = 3.312 \, V$

$3 \, \Omega$ resistor: $1.103 \times 3 = 3.309 \, V$

(c)

$$P = I^2 R_3 = (1.517)^2 \times 3 = 6.9 \, W$$

$$P = I^2 R_2 = (1.517)^2 \times 2 = 4.6 \, W$$

$$P = I_1^2 \times R_2 = (0.759)^2 \times 2 = 1.15 \, W$$

$$P = I_2^2 \times R_3 = (0.506)^2 \times 3 = 0.77 \, W$$

$$P = I_3^2 R_6 = (0.252)^2 \times 6 = 0.38 \, W$$

$$P = I_4^2 R_8 = (0.414)^2 \times 8 = 1.37 \, W$$

$$P = I_5^2 R_3 = (1.103)^2 \times 3 = 3.65 \, W$$

$$P = I^2 R_5 = (1.517)^2 \times 5 = 11.5 \, W$$

Example 1.9

A television receiver extracts proportions of red, blue and green colours from the colour signal by means of a simple resistor matrix as shown in Fig. 1.18. The output difference signals from the matrix are then applied to the colour

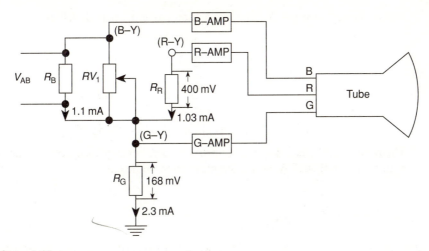

Figure 1.18

amplifiers and the television tube. The voltages and currents required for correct colour distribution on the screen are given. Determine:

(a) the setting of RV_1 under these conditions;
(b) the resistance range and rating of RV_1 if the current through R_G is required to vary from 2.5 to 3.2 mA, assuming the currents through R_B and R_R have to remain consistent.

Solution

(a) As R_B and RV are in parallel, the effective current taken by them is

$$2.3 - 1.03 = 1.27\,\text{mA}$$

∴ the current through RV_1 is

$$1.27 - 1.1 = 0.16\,\text{mA}$$

hence

$$RV_1 = \frac{V}{I} = \frac{280 \times 10^{-3}}{10^{-3} \times 0.16} = 1.75\,\text{k}\Omega$$

(b) When the R_G current is 2.5 mA then current through RV_1

$$2.5 - 1.03 = 1.47\,\text{mA}$$

$$\therefore\ 1.47 - 1.1 = 0.37\,\text{mA}$$

$$\therefore\ RV_1 = \frac{280 \times 10^{-3}}{0.37 \times 10^{-3}} = 756.75\,\Omega \qquad (1.10)$$

$$P = I^2 RV_1 = \left(\frac{0.37}{10^3}\right)^2 \times RV_1 = 103.6\,\mu\text{W}$$

When current through R_G is 3.2 mA then current through RV is

$$3.2 - 1.03 = 2.17\,\text{mA}$$

$$\therefore\ 2.17 - 1.1 = 1.07\,\text{mA}$$

$$RV_1 = \frac{280 \times 10^{-3}}{1.07 \times 10^{-3}} = 261.68\,\Omega$$

$$P = I^2 RV_1 = (1.07 \times 10^3)^2 \times 261.68 = 299.6\,\mu\text{W}$$

The range of RV_1 is 261–757 Ω approximately, hence a 1 kΩ potentiometer with a rating of 0.25 W would be more than sufficient.

1.9 Kirchhoff's laws

Any discussion on d.c. circuit analysis would not be complete without at least an understanding of what Kirchhoff's laws tell us:

1. *Kirchhoff's first law (current law)*. This is simply a statement of the principle of conservation of charge and can be stated as the sum of the currents flowing into a junction is equal to the sum of the currents flowing out of the junction.
2. *Kirchhoff's second law (voltage law)*. This law states that the algebraic sum of the p.d.s round any closed loop in a circuit equals zero. Remember that potential can be positive or negative, hence p.d.s must be given a sign. This has already been shown in Fig. 1.5.

These laws have been used in a limited form for the solutions of some of the examples already shown. Analytical solutions tend to be cumbersome and for this reason other theorems are used. However, an example is included here for completeness.

Example 1.10

For the network shown in Fig. 1.19 determine the current in each branch and the voltage drop across the 5 Ω resistor. The assumed current directions are as

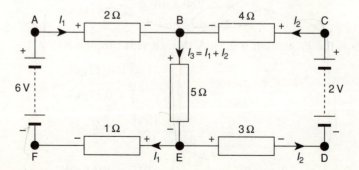

Figure 1.19

shown. Note if the answers for the currents are negative then this means the assumed current direction was wrong. The potentials are marked according to conventional current flow.

By Kirchhoff's first law at B

$$I_1 + I_2 - I_3 = 0 \qquad (1.11)$$

Thus

$$I_3 = I_1 + I_2 \qquad (1.12)$$

This reduces the number of unknown currents to two. Kirchhoff's current law tells us that at junction E, $I_1 + I_2$ flows to the junction, thus $I_1 + I_2$ must flow from the junction as shown. As there are two unknown currents we require two equations and any two loops.

Take loop ABEFA
Going round the loop in a clockwise direction using Kirchhoff's voltage law, the voltage equation becomes

$$V_{AB} + V_{BE} + V_{EF} + V_{FA} = 0$$

Thus

$$2I_1 + 5(I_1 + I_2) + 1I_1 - 6 = 0 \qquad (1.13)$$

gives

$$8I_1 + 5I_2 = 6$$

Take loop BCDEB
Going round in a clockwise direction, the voltage equation becomes

$$V_{BC} + V_{CD} + V_{DE} + V_{EB} = 0$$

$$-4I_2 + 2 - 3I_2 - 5(I_1 + I_2) = 0 \qquad (1.14)$$

This gives

$$-5I_1 - 12I_2 = -2$$

The two simultaneous equations are (1.13) and (1.14)

$$8I_1 + 5I_2 = 6$$

$$-5I_1 - 12I_2 = -2$$

These may be solved in two ways.

Method 1
Multiply (1.13) × 12:

$$96I_1 + 60I_2 = 72 \qquad (1.15)$$

Multiply (1.14) × 5:

$$-25I_1 - 60I_2 = -10 \qquad (1.16)$$

(1.15) + (1.16) gives

$$71I_1 = 62$$

$$I_1 = 0.873\,\text{A}$$

Substitute in (1.13):

$$8 \times 0.873 + 5I_2 = 6$$

$$\therefore I_2 = -0.197\,\text{A}$$

The negative sign indicates that I_2 flows in the opposite direction to that assumed.

Method 2

Rearranging equation (1.13) gives

$$I_2 = \frac{6 - 8I_1}{5} \qquad\qquad (1.17)$$

Substitute (1.17) in (1.14):

$$-5I_1 - 12\left(\frac{6 - 8I_1}{5}\right) = -2$$

$$-5I_1 - 14.4 + 19.2I_1 = -2$$

$$14.2I_1 = 12.4$$

$$\therefore I_1 = 0.873\,\text{A}$$

Substitute in (1.17):

$$I_2 = \frac{6 - (8 \times 0.873)}{5}$$

$$I_2 = -0.197\,\text{A}$$

Finally the current through the $5\,\Omega$ resistor is

$$I_1 + I_2 = 0.873 - 0.197 = 0.676\,\text{A}$$

Voltage drop across the $5\,\Omega$ resistor is

$$5 \times 0.676 = 3.38\,\text{V}$$

Note that $V_{\text{BE}} = +3.38\,\text{V}$ and $V_{\text{EB}} = -3.38\,\text{V}$

1.10 The Wheatstone bridge

This type of network is a commonly used network in solving earth faults in cables. It is also used in monitor circuits where transducers such as optical, heat and pressure sensors form part of the bridge. Figure 1.20 shows the principle of operation. If all the resistances have values such that no current flows through the monitor, the bridge is said to be balanced. At

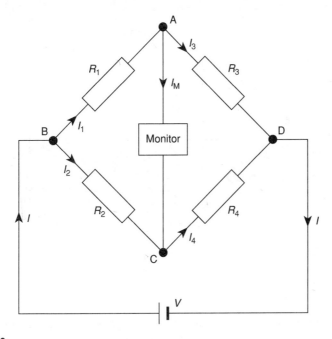

Figure 1.20

balance

$$I_M = 0 \quad \text{and} \quad \text{Potential A} = \text{Potential C}$$

$$\therefore \quad V_{BA} = V_{BC} \quad \text{and} \quad V_{DA} = V_{DC}$$

$$I_1 R_1 = I_2 R_2 \quad \text{and} \quad I_3 R_3 = I_4 R_4$$

Dividing gives

$$\frac{I_1 R_1}{I_3 R_3} = \frac{I_2 R_2}{I_4 R_4}$$

But because $I_M = 0$

$$I_1 = I_3 \quad \text{and} \quad I_2 = I_4$$

$$\therefore \quad \frac{R_1}{R_3} = \frac{R_2}{R_4} \tag{1.18}$$

Example 1.11

A test is used to find the position of an earth fault on an 80 km length of twin cable. The resistance is 20 Ω/km in each cable. The circuit is shown in Fig. 1.21. Assume balance occurs when $R_1 = R_2$ and $RV = 2\,k\Omega$.

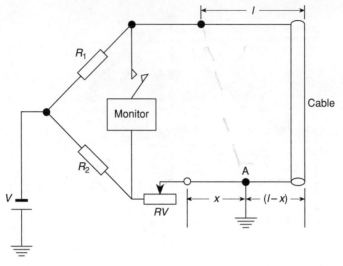

Figure 1.21

This circuit is based on the Wheatstone bridge. From equation (1.18)

$$\frac{R_1}{R_3} = \frac{R_2}{R_4}$$

If r is the resistance per kilometre of the line then at balance

$$\frac{R_1}{[l + (l - x)]r} = \frac{R_2}{RV + xr}$$

where

$$[l + (l - x)]r = R_3 \quad \text{and} \quad (RV + xr) = R_4$$
$$\therefore \ R_1 RV + R_1 xr = 2R_2 rl - R_2 rx$$

Arranging gives

$$x = \frac{2R_2 rl - R_1 RV}{r(R_1 + R_2)}$$
$$= \frac{2 \times 500 \times 20 \times 80 - 500 \times 2000}{20 \times 1000}$$
$$= \frac{15 \times 10^5}{2 \times 10^4} = 75 \,\text{km}$$

The earth fault is 75 km along the cable which is connected to the variable resistor RV.

Example 1.12

A Wheatstone bridge is used as a temperature sensing circuit together with a differential amplifier (Fig. 1.22). Determine:

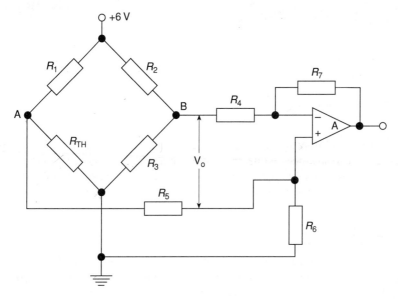

Figure 1.22

(a) the resistance of the thermistor at the calibration temperature;
(b) the voltage which will be applied to the amplifier if a change of $5\,\Omega$ occurs in the thermistor's resistance.

Solution

(a) At the calibration temperature the bridge will be balanced when

$$\frac{R_1}{R_2} = \frac{R_{TH}}{R_3}$$

$$\therefore\ R_{TH} = \frac{R_3 R_1}{R_2}$$

$$= \frac{680 \times 1200}{2200} = 371\,\Omega$$

(b) The output voltage will be the difference in potential between the points A and B. Using voltage division gives the output voltage (V_o) at the calibration frequency:

$$V_o = V_s[R_{TH}/(R_1 + R_{TH}) - R_3/(R_2 + R_3)]$$

For a resistance charge of $r = 5\,\Omega$ in the thermistor,

$$V_o = V_s[(R_{TH} + r)/[R_1 + (r + R_{TH})] - R_3/(R_2 + R_3)]$$

$$= 6[(371 + 5)/[1200 + (5 + 371)] - 680/(2200 + 680)]$$

$$= 6 \times \{376/1576 - 680/2880\} = 0.018\,V$$

1.11 Attenuator networks

Attenuators are resistive networks and are used in many applications where a source has to be matched to a load. They are generally designed in two configurations:

1. Symmetrical attenuators
2. Non-symmetrical attenuators.

Symmetrical attenuators are used where the source and load have equal characteristic resistances while the non-symmetrical attenuator is used where there is a mismatch between source and load. Two types are of particular interest in d.c. networks: the unsymmetrical L and symmetrical T types.

Example 1.13

A non-symmetrical attenuator is shown in Fig. 1.23 between a source and a load. The source and load characteristic impedances are respectively 100 and $60\,\Omega$. Determine the current through the load and across it.

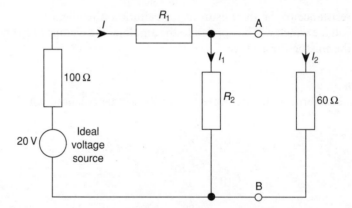

Figure 1.23

Solution
The purpose of this exercise is to make the load 'see' a resistance of $100\,\Omega$ at the terminals AB. Hence a balance equation using voltage division is used:

$$100 = R_1 + 60R_2/(60 + R_2)$$

$$(60 + R_2)100 = R_1(60 + R_2) + 60R_2$$

$$6 \times 10^3 + 100R_2 = 60R_1 + R_1R_2 + 60R_2$$

$$6 \times 10^3 = 60R_1 + R_1R_2 - 40R_2$$

(1.19)

Also

$$60 = \frac{R_2(R_1 + 100)}{R_2 + R_1 + 100}$$

$$6 \times 10^3 + 60R_2 + 60R_1 = R_2R_1 + 100R_2$$

$$6 \times 10^3 = R_2R_1 + 100R_2 - 60R_2 - 60R_1 \qquad (1.20)$$

$$= 40R_2 + R_2R_1 - 60R_1$$

Subtracting equations (1.19) and (1.20) gives

$$0 = 120R_1 - 80R_2$$

$$\therefore \frac{R_2}{R_1} = \frac{120}{80} \qquad (1.21)$$

$$\therefore R_2 = 1.5R_1$$

Adding equations (1.19) and (1.20) gives

$$12 \times 10^3 = 2R_1R_2$$

$$\therefore R_1 = \frac{6 \times 10^3}{R_2} \qquad (1.22)$$

Substitute equation (1.22) into (1.21):

$$R_2 = 1.5\frac{[6 \times 10^3]}{R_2}$$

$$R_2^2 = 9 \times 10^3 \qquad (1.23)$$

$$R_2 = 95\,\Omega$$

Substitute equation (1.23) into (1.21):

$$95 = 1.5R_1$$

$$\therefore R_1 = 63\,\Omega$$

The current flowing in the load is by current division

$$I_2 = \left(\frac{95}{95 + 60}\right)I \qquad (1.24)$$

Since the equivalent resistance of the network is

$$37 + 100 + R_1 = 37 + 100 + 63 = 200$$

then

$$I = \frac{20}{200} = 0.1\,\text{A}$$

Hence equation (1.24) becomes

$$I_2 = \left(\frac{95}{95 + 60}\right)0.1 = 0.06\,\text{A}$$

Example 1.14

Design a symmetrical T-attenuator having $R = 650\,\Omega$ and an attenuation of 15 dB (Fig. 1.24).

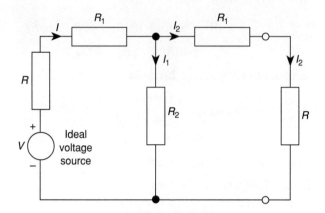

Figure 1.24

Solution
Note 15 dB = 5.62 = attenuation constant. By current division

$$I_2 = \left(\frac{R_2}{R_1 + R_2 + R}\right)I$$

$$\therefore \frac{I}{I_2} = \frac{(R_1 + R_2 + R)}{R_2} = A \tag{1.25}$$

where A is the attenuation constant

$$R = R_1 + R_2(R_1 + R)/R_1 + R_2 + R$$
$$= R_1 + (R_1 + R_0)/A$$
$$RA = R_1A + R_1 + R$$

Separating terms gives

$$R(A - 1) = R_1(A + 1)$$

$$\therefore \ R_1 = R(A - 1)/(A + 1) \tag{1.26}$$

From equation (1.25)

$$R_2A = R_1 + R_2 + R$$
$$R_2(A - 1) = R_1 + R$$
$$= R(A - 1)/(A + 1) + R \tag{1.27}$$
$$R_2(A - 1)(A + 1) = R(A - 1) + R(A + 1) = 2RA$$
$$\therefore \ R_2 = 2RA/(A^2 - 1)$$

From equation (1.26)

$$R_1 = R(A-1)/(A+1)$$
$$= 650(5.62-1)/(5.62+1)$$
$$\therefore \ R_1 = \frac{2777}{4.62} = 601\,\Omega$$

From equation (1.27)

$$R_2 = 2RA/(A^2-1)$$
$$= \frac{2 \times 650 \times 5.67}{(5.62^2-1)}$$
$$= \frac{7306}{30.6} = 239\,\Omega$$

Example 1.15

Design a symmetrical T-attenuator which will provide an attenuation of 20 dB between the cables (20 dB = 10) (Fig. 1.25).

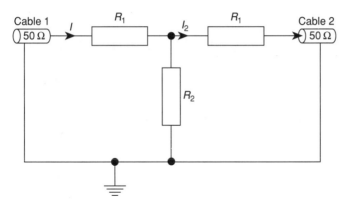

Figure 1.25

Solution
This example is solved in a similar fashion to Example 1.14 using equations (1.26) and (1.27):

$$R_1 = 50(10-1)/(10+1)$$
$$\therefore \ R_1 = 41\,\Omega$$
$$R_2 = 2 \times 50 \times 10/(100-1) = 10\,\Omega$$

1.12 Equivalent circuits

It has already been seen in this chapter that resistor networks can be replaced by an equivalent resistor. Examples 1.7 and 1.8 illustrate this. This means that as far as a supply is concerned any network of resistors can be replaced by a single equivalent resistor. Hence, no matter how complex a circuit is, it may be replaced by a constant voltage source in series with a resistance or a constant current source in parallel with a resistance.

1.13 Thévenin's theorem

The constant voltage source is called the Thévenin voltage and the series resistance is called the Thévenin resistance. This principle will now be examined in a detailed example.

Example 1.16

In this problem it is necessary to calculate the current through and the voltage across the $2\,\Omega$ load (Fig. 1.26). As far as the load is concerned, the section enclosed in the dotted box in Fig. 1.26 can be replaced by a constant voltage source in series with a resistance. This is the Thévenin equivalent circuit (Fig. 1.27).

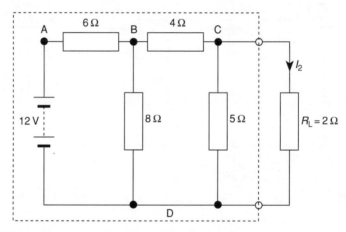

Figure 1.26

To reach this stage R_T and V_T must be found. This can be done as follows:

(a) Disconnect the load.
(b) Calculate the open circuit voltage. This is the Thévenin voltage V_T.
(c) With the load still disconnected, short all the supplies, retaining any internal resistance which they may have, and calculate the resistance looking into the load terminals C and D. This is the Thévenin resistance R_T.
(d) Connect the load to the Thévenin equivalent circuit and calculate the current and voltage for the load.

Applying steps (a) and (b) gives Fig. 1.27.

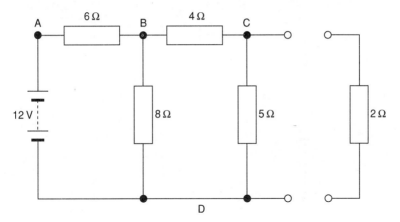

Figure 1.27

The 4 Ω and 5 Ω resistors are in series and can be replaced by 9 Ω (Fig. 1.28). The 8 Ω and 9 Ω resistors are in parallel and can be replaced by a resistor of 4.24 Ω (Fig. 1.29). The total resistance across the supply is 10.24. Current taken from supply is

$$I = \frac{12}{10.24} = 1.17\,\text{A}$$

Referring back to Fig. 1.28, this current will split at **B**. Hence by current division the current through the 9 Ω resistor is

$$\left(\frac{8}{8+9}\right) \times 1.17 = 0.55\,\text{A}$$

Transferring this back to Fig. 1.27, the current through the 4 Ω and 5 Ω series resistors is 0.53. The voltage across the 5 Ω resistor is

$$V_{\text{CD}} = 0.55 \times 5 = 2.75\,\text{V}$$

This is the Thévenin voltage V_{T}.

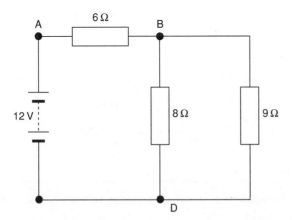

Figure 1.28

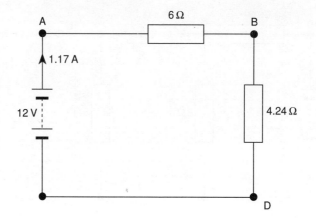

Figure 1.29

Applying step (c) will give Fig. 1.30. As the $6\,\Omega$ and $8\,\Omega$ resistors are in parallel this gives $3.43\,\Omega$. Figure 1.31 follows. The $3.43\,\Omega$ and $4\,\Omega$ resistors are in series and this gives $7.43\,\Omega$. Figure 1.32 follows. The $7.43\,\Omega$ and $5\,\Omega$ resistors are in parallel and can be replaced by an equivalent resistor of $2.99\,\Omega$. This is the Thévenin resistance R_T (Fig. 1.33). Step (d) now follows.

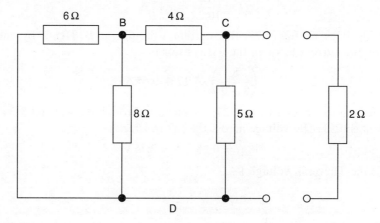

Figure 1.30

The current through the load is

$$I = \frac{2.75}{4.99} = 0.55\,\text{A}$$

$$V_{CD} = 0.55 \times 2 = 1.1\,\text{V}$$

Example 1.17

An operational amplifier incorporates a ladder network at its input as shown in Fig. 1.34. Determine the output voltage of the amplifier if the gain of the amplifier is 20 dB (10). Input resistance to the amp. is $10\,\text{M}\Omega$.

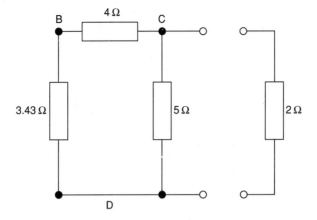

Figure 1.31

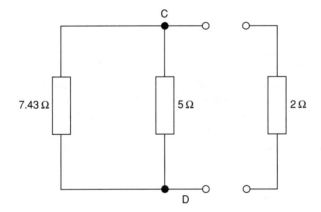

Figure 1.32

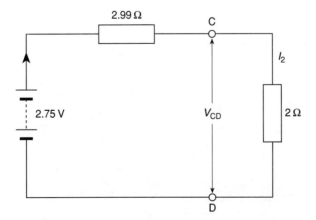

Figure 1.33

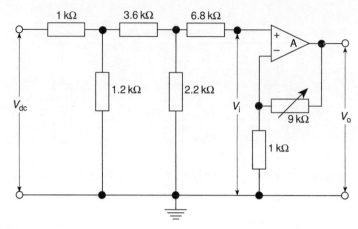

Figure 1.34

Solution

Short-circuiting the supply and open circuiting the input terminals to the amplifier will enable the Thévenin resistance to be calculated as shown by Figs 1.35–1.38.

Figure 1.39 enables the Thévenin voltage to be calculated. According to Ohm's law an ideal voltmeter with a very high resistance would not take any current at the terminal. Most industrial voltmeters have a very high resistance in order to avoid disturbing the circuit so the voltage measured at CD will be the same as at AB, i.e. as no current flows through the 6.8 kΩ

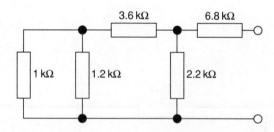

Figure 1.35

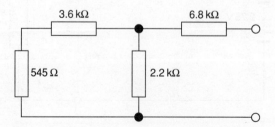

Figure 1.36

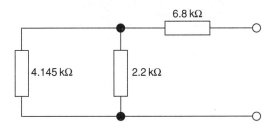

Figure 1.37

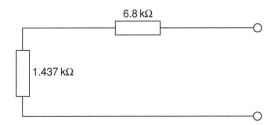

Figure 1.38

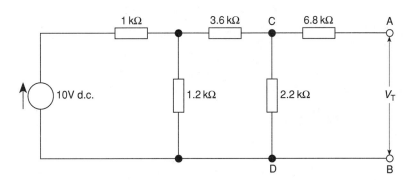

Figure 1.39

resistor there is no voltage drop. The $3.6\,\mathrm{k\Omega}$ resistor is in parallel with the $2.2\,\mathrm{k\Omega}$ resistor.

$$\therefore\ \frac{2.2 \times 3.6}{2.2 + 3.6} = 1.37\,\mathrm{k\Omega}$$

The circuit now reduces to Fig. 1.40. This leads to Fig. 1.41.

$$\therefore\ I_\mathrm{T} = \frac{10}{1400} = 7.142\,\mathrm{mA}$$

Referring to Fig. 1.40, the current through the $1.37\,\mathrm{k\Omega}$ combination is

$$\left(\frac{1.2}{1.2 + 1.37}\right) \times 7.142 = 3.33\,\mathrm{mA}$$

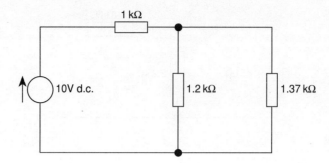

Figure 1.40

Referring to Fig. 1.39, the current through the 2.2 kΩ resistor is

$$\left(\frac{3.6}{3.6+2.2}\right) \times 3.33 = 2.07\,\text{mA}$$

$$\therefore\ V_T = 2.07 \times 10^{-3} \times 2.2 \times 10^3 = 4.554\,\text{V}$$

The equivalent Thévenin circuit with 10 MΩ load gives Fig. 1.42.

$$\therefore\ I_L = \frac{V}{R} = \frac{4.554}{1\,008\,237} = 4.5\,\mu\text{A}$$

$$\therefore\ V_L = 4.5 \times 10^{-6} \times 10^6 = 4.5\,\text{V}$$

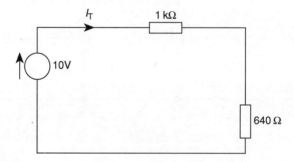

Figure 1.41

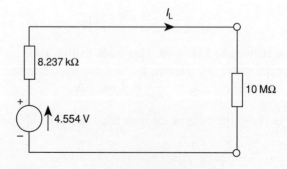

Figure 1.42

The output voltage from the operational amplifier is

$$V_o = 1.1 \times 10 = 11\,\text{V}$$

1.14 Norton's theorem

As an alternative to Thévenin's theorem, Norton's theorem states that any circuit supplying a load may be replaced by a constant current source in parallel with a resistance. The constant current source is equal to the short circuit obtained by shorting the load. The resistance is the resistance seen when looking into the load terminals with the load disconnected. The Norton resistance is therefore exactly the same as the Thévenin resistance, but is connected in parallel.

Example 1.18

Figure 1.43 is the same circuit as Example 1.15, but this time we will find the current through the loop and the voltage across it by Norton's theorem. As far as the load is concerned, the section enclosed in the box can be replaced by a constant current source in parallel with a resistance. Figure 1.44 shows the equivalent Norton circuit.

Initially it is necessary to calculate the short circuit current I_{SC} and then the Norton resistance (R_N). The following steps are required:

(a) Short circuit the load.
(b) Calculate the short circuit current.
(c) Disconnect the load, short the supplies (retaining any internal resistance) and calculate the resistance looking into the load terminals CD. This is the Norton resistance.
(d) Connect the load to the Norton equivalent circuit and calculate the current and voltage for the load.

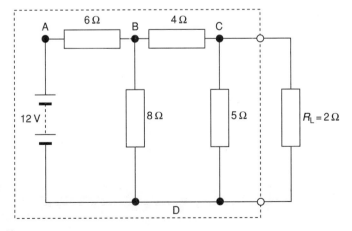

Figure 1.43

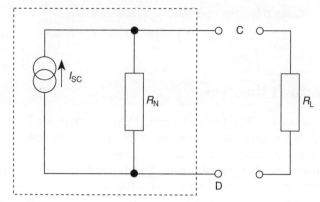

Figure 1.44

Applying (a) and (b) to Fig. 1.43 gives Fig. 1.45. Note that the short circuit across the load terminals shorts the $2\,\Omega$ load resistor and the $5\,\Omega$ resistor, so I_{SC} is the current flowing in the $4\,\Omega$ resistor. The $8\,\Omega$ and $4\,\Omega$ resistors are in parallel and can be replaced by a $2.67\,\Omega$ resistor (Fig. 1.46).

The current taken from the supply is

$$\frac{12}{(6+2.67)} = 1.38\,\text{A}$$

Referring to Fig. 1.43, this current will split at B to give a short circuit current I_{SC} flowing through the $4\,\Omega$ resistor:

$$\therefore\ I_{SC} = \frac{8}{(4+8)} \times 1.38 = 0.92\,\text{A}$$

Applying step (c) gives the same answer as for the Thévenin circuit, i.e. $R_N = 2.99\,\Omega$. The Norton equivalent circuit now becomes that shown in Fig. 1.47. Applying step (d) to Fig. 1.47 shows that I_{SC} splits at C. The current

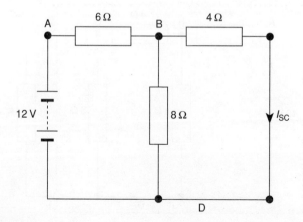

Figure 1.45

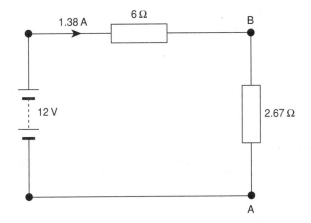

Figure 1.46

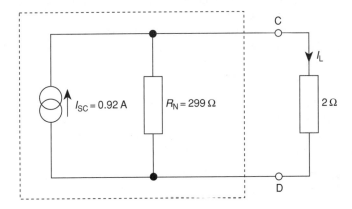

Figure 1.47

flowing through the $2\,\Omega$ load can be found as follows:

$$I_{\mathrm{L}} = \left(\frac{2.99}{2.99 + 2}\right) \times 0.92 = 0.55\,\mathrm{A}$$

$\therefore$ Voltage across the load is

$$V_{\mathrm{CD}} = 2 \times 0.55 = 1.1\,\mathrm{V}$$

These answers are identical to those obtained with the Thévenin network.

1.15 Further problems

1. The circuit shown in Fig. 1.48 is connected with wire having a resistance of $1.5\,\Omega/\mathrm{m}$. If each lamp is rated at $500\,\mathrm{W}$ and the maximum permitted voltage drop on the lines is $16\,\mathrm{V}$, determine:

(a) if this wiring arrangement is suitable;
(b) if the lamps in bank 2 are delivering full light.

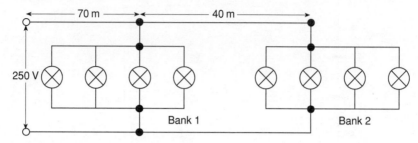

Figure 1.48

Answer: 14.4 V

2. A circuit network consists of two parallel resistors having resistances of 20 and 30 Ω respectively, connected in series with a 15 Ω heater element. If the current through the heater is 3 A, find:
   (a) the current through the 20 and 30 Ω resistors;
   (b) the supply voltage;
   (c) the total power.
   Answer: 1.8 A; 1.2 A; 81 V; 243 W

3. Four resistors each of 150 Ω are connected in parallel and then joined in series with a heater element of 15 Ω. If the supply voltage is 240 V calculate the heat dissipated by the heater each minute.
   Answer: 65.8 kJ

4. An accumulator undergoes a voltmeter test. On open circuit the voltmeter reads 13 V and this falls to 12.5 V when a 0.5 Ω resistor is connected across the terminals of the accumulator. Calculate:
   (a) the current;
   (b) the internal resistance;
   (c) the power generated in the accumulator.
   Answer: 25 A; 0.02 Ω; 0.5 W

5. A two-wire distributor is shown in Fig. 1.49. The distributed load points are indicated at A, B, C and D and the cable used has a resistance of 0.03 Ω/ 100 m. Calculate:

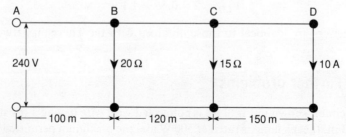

Figure 1.49

(a) the voltage at the load points B, C and D;
(b) the total power loss in the distributor.
Answer: 237.3 V; 235.19 V; 235 V; 175.5 W

6. The circuit shown in Fig. 1.50 is used to charge two banks of cells having different ratings. B1 consists of a battery of 50 cells each of 2 V and internal resistance 0.1 Ω; B2 consists of a battery of 20 cells each of 1.5 V and internal resistance 0.15 Ω. If the generator has an open circuit EMF of 150 V and an armature resistance of 0.25 Ω, determine the rates of charge if R_1 and R_2 are set at 2.5 and 69 Ω respectively.
Answer: 2 A; 1.5 A

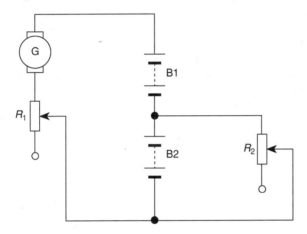

Figure 1.50

7. A resistor ladder network is required to produce a 2 V output as part of an interface circuit consisting of an analogue-to-digital converter. The circuit is shown in Fig. 1.51.

    Show, using Thévenin's theorem, that this network is suitable for the specification.

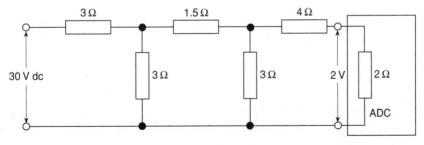

Figure 1.51

8. In Fig. 1.52 determine the voltage at the output of the network when the switch S is open, the voltage across the control unit and the current through its internal resistance when the switch is closed.
Answer: 4.01 V; 2.28 V; 431 mA

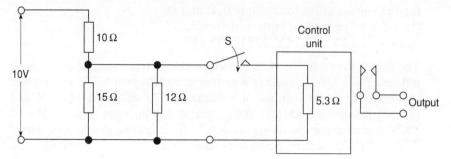

Figure 1.52

9. In the sensing circuit shown in Fig. 1.53, $R_1 = 2.2\,\text{k}\Omega$, $R_2 = 3.2\,\text{k}\Omega$ and $R_3 = 820\,\Omega$. Determine:
   (a) the resistance of the thermistor at the calibration temperature;
   (b) the voltage applied to the amplifier if a change of $3\,\Omega$ occurs in the thermistor's resistance.
   Answer: $564\,\Omega$; $10\,\text{mV}$

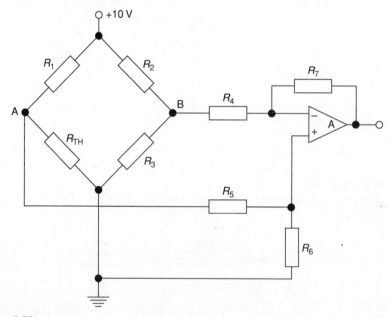

Figure 1.53

10. An earth fault occurs in a 100 km length of twin cable. Determine where this fault occurs if the resistance is $12\,\Omega/\text{km}$ in each cable, $R_1 = R_2 = 300\,\Omega$ and $RV = 1.5\,\text{k}\Omega$.
    Answer: 37 km

11. A non-symmetrical attenuator is shown in Fig. 1.54. It is placed between a source and a load as shown. Determine the current through the load and the voltage across it.
    Answer: 0.06 A; 7.5 V

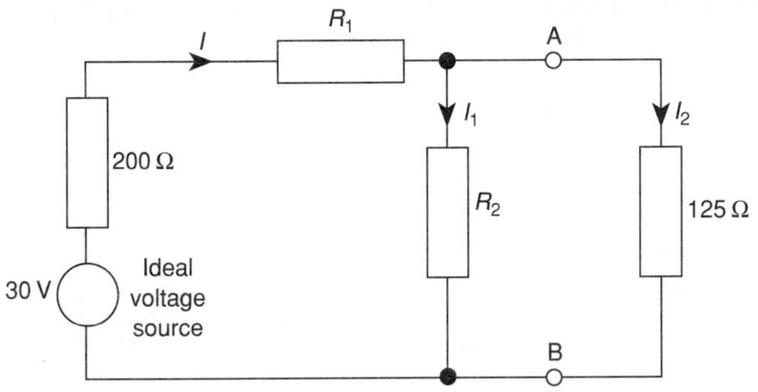

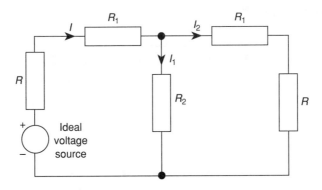

Figure 1.54

12. Design a T-attenuator having $R = 500\,\Omega$ and an attenuation of 10 dB (Fig. 1.55).
    Answer: $R_1 = 260\,\Omega$; $R_2 = 352\,\Omega$

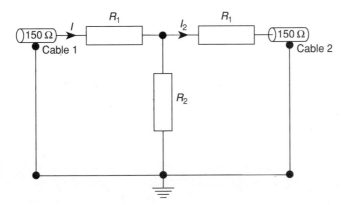

Figure 1.55

13. Design a symmetrical T-attenuator which will provide an attenuation of 32 dB between the cables (Fig. 1.56).
    Answer: $R_1 = 143\,\Omega$; $R_2 = 7.5\,\Omega$

Figure 1.56

14. Determine the voltage across the load R_L by using Norton's and Thévenin's theorems (Fig. 1.57).

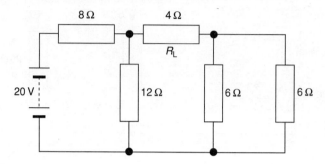

Figure 1.57

Answer: 4.07 V

15. Determine the voltage across the load R_L by using Norton's and Thévenin's theorems (Fig. 1.58).

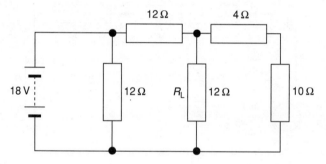

Figure 1.58

Answer: 6.3 V

2
___

Capacitive networks

2.1 Introduction

Figure 2.1 shows a simple power supply which incorporates a bridge rectifier, a power-on LED, filter capacitor (C), zener regulator and an output regulator which controls the output voltage for varying loads.

One of the problems which is often met with in this circuit is the correct selection of the filter capacitor which has to ensure low ripple content at the output. This in turn depends on the CR time constant of the zener diode (D_2) and filter capacitor combination. These problems and others will be examined in this chapter.

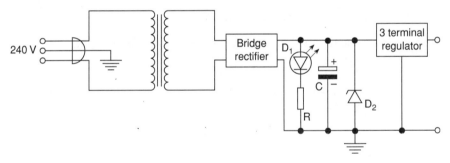

Figure 2.1

2.2 Electric field intensity

Most electrical and electronic circuits consist of active and passive components, active components being those which produce energy, and passive components being those which do not. Capacitance may be said to be both passive and active, in that a capacitor has the ability to store energy even when disconnected from a circuit.

From Chapter 1 it was seen that

$$I = \frac{Q}{t} \quad \text{or} \quad Q = It \tag{2.1}$$

and when work is done on electrical charge an electrical potential energy difference is created. In general

$$E = QV \qquad (2.2)$$

where E is the energy produced in joules and V is measured in volts. This basic principle is used in many applications such as electrostatic deflection, CR networks, filtering and capacitive transducers.

Before tackling situations where capacitance is involved, the basic physical concepts of the electrostatic field should be clarified and one particular application will be considered in section 2.3. Everyone is familiar with electric discharge, whether it is static charge produced in clothing during dry weather or the more dramatic manifestation during electrical storms. In both cases, however, the cause is the same, namely the establishment of two points having different potentials. Generally we consider these points as having negative and positive potentials.

Consider two parallel plates connected to a battery with a potential difference of V volts (Fig. 2.2). The upper plate is given a positive charge while the bottom plate is given a negative charge due to electrostatic conduction or transfer of charge. The plates therefore have a p.d. between them. Since the plates are parallel the electric field below them is uniform and if a charge of $+q$ coulombs is placed between the plates it will experience a force of F newtons (N) downwards. If the charge has to reach the upper plate a force F will have to be applied upwards. Hence the work done in taking the charge from the lower plate to the upper plate is

$$\text{Force} \times \text{distance} = F \times d$$

Also the charge gains an electrical potential energy difference given by

$$F \times d = qV \qquad (2.3)$$

If an electronic charge is involved this is given as eV, which is a common unit of energy called the electron-volt or

$$\frac{F}{q} = \frac{V}{d} \qquad (2.4)$$

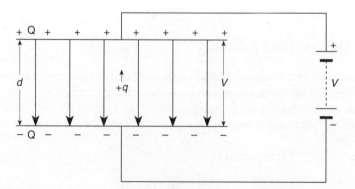

Figure 2.2

In Newtonian mechanics the strength of a gravitational field is measured by a force acting on a mass, i.e. newton/kg. Equation (2.3) is a measure of the force acting on a unit. This is called the electric field intensity, but it is not easy in practice to measure the force on a charge. For this reason the electric field intensity is calculated by using

$$\xi = \frac{V}{d} \text{ volts m}^{-1} \tag{2.5}$$

where d is the distance between the plates and V is the voltage between them. This is sometimes referred to as the potential gradient.

2.3 The cathode ray oscilloscope

It is informative to examine the electron physics involved in the cathode ray tube shown in Fig. 2.3. In this diagram the electron beam originates from the cathode and is focused by means of the electrostatic field between the focusing and accelerating anodes which are at high potential. The accelerated electrons pass between two pairs of deflecting plates which cause the beam to be deflected vertically and horizontally.

The electron gun causes a velocity (v_g) to be imparted to the electrons. In order to calculate this energy the conservation of energy is used, i.e. the energy due to the electrons being evaporated from the cathode is equal to the energy after acceleration due to the anode:

$$\tfrac{1}{2}mv_k^2 + ev_k = \tfrac{1}{2}m_a v_a^2 + ev_a \tag{2.6}$$

This comes about due to equation (2.3) in which energy is imparted to the electronic charge. Although the electrons have a velocity (v_k) when they leave the cathode, this is so small compared with their final velocity (v_a) that it can be neglected so that

$$v_a = \sqrt{\frac{2e(v_k - v_a)}{m}} = \sqrt{\frac{2eV_1}{m}} \tag{2.7}$$

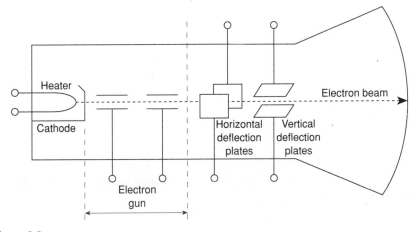

Figure 2.3

The potential difference between anode and cathode is given by V_1 while m is the mass of the electron and e is the electronic charge. The velocity given in equation (2.7) is the velocity at which the electron beam enters the deflection plates as shown in Fig. 2.4.

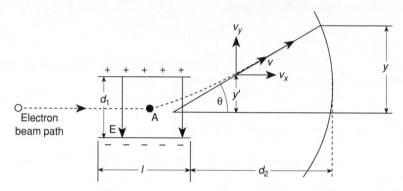

Figure 2.4

The electric field intensity is given by equation (2.5):

$$E = \frac{V_2}{d_1} \qquad (2.8)$$

where V_2 is the potential difference between the plates.

A constant upward force eV acts on the electrons and their upward acceleration is

$$a_y = \frac{eE}{m} \qquad (2.9)$$

As the horizontal velocity remains constant the time required to travel the length (l) of the plate is

$$t = \frac{l}{v_x} \qquad (2.10)$$

The upward velocity in this time is

$$v_y = a_y t \qquad (2.11)$$

and the upward displacement is, when (2.9) and (2.10) are substituted:

$$y = \frac{1}{2} a_y t^2 = \frac{eEl^2}{2mv_x^2} \qquad (2.12)$$

The angle of deflection is

$$\tan \theta = \frac{v_y}{v_x} \qquad (2.13)$$

The beam then passes to the screen in a straight line. As can be seen from Fig. 2.4, if the trajectory is projected backwards to midway between the plates then

$$\tan \theta = \frac{y}{d_2 + (d_1/2)}$$

However, as the length $d_1 \ll d_2$, then

$$\tan\theta = \frac{y}{d_2} \qquad (2.14)$$

Hence

$$\frac{v_y}{v_x} = \frac{y}{d_1} = \tan\theta \qquad (2.15)$$

All these parameters are measurable and it is therefore possible to set up the deflection plate system for a constant value of V_1.

Example 2.1

A cathode ray tube has to be designed with the following parameters in order to allow maximum screen height (Fig. 2.5):

Length of plates	$l = 4\,\text{cm}$
Distance to screen	$d_2 = 12\,\text{cm}$
Accelerating voltage	$V_1 = 182\,\text{V}$
Distance between plates	$d_1 = 2\,\text{cm}$

Determine the screen height if the maximum voltage between the vertical plate is 120 V.

Solution

$$v_x = \sqrt{\frac{2eV_1}{m}} = \sqrt{\frac{2 \times 1.6 \times 10^{-19} \times 182}{9.11 \times 10^{-31}}} = 8 \times 10^6 \, \text{m s}^{-1}$$

From (3.8)

$$E = \frac{V_2}{d_1} = \frac{120}{2} = 6 \times 10^3 \, \text{V m}^{-1}$$

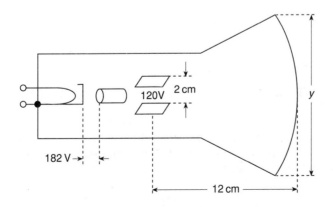

Figure 2.5

hence from (2.12)

$$y' = \frac{eEl^2}{2mv_x^2}$$

$$= \frac{1.6 \times 10^{-19} \times 6 \times 10^3 \times 4 \times 10^{-2}}{2(9.11 \times 10^{-31})(8 \times 10^6)} = 1.32\,\text{cm}$$

Also from (2.10)

$$t = \frac{1}{v_x} = \frac{4 \times 10^{-2}}{8 \times 10^6} = 5\,\text{ns}$$

Equation (2.9) gives

$$a_y = \frac{eE}{m} = \frac{1.6 \times 10^{-19} \times 6 \times 10^3}{9.11 \times 10^{-31}}$$

$$= 1.054 \times 10^{15}\,\text{m s}^{-1}$$

and from (2.11)

$$v_y = a_y t = 1.054 \times 10^{15} \times 5 \times 10^{-9}$$

$$= 5.27 \times 10^6\,\text{m s}^{-1}$$

From (2.13)

$$\tan\theta = \frac{v_y}{v_x} = \frac{5.27 \times 10^6}{8 \times 10^6} = 0.6587$$

Also (2.15) gives

$$\frac{v_y}{v_x} = \frac{y}{d_2} = \tan\theta$$

$$\therefore\ y = \tan\theta d_2 = 12 \times 0.6587 = 7.9\,\text{cm}$$

$\therefore$ The total deflection is $7.9 + 1.32 = 8.22\,\text{cm}$ and a 9 cm tube would be appropriate.

2.4 Capacitance

As was seen from section 2.2, two plates with charge on them cause an electric field to be established between them. The values of the p.d. between the plates for different values of charge on the plate are compared and it can be shown that, for a particular capacitor,

$$\frac{Q}{V} = \text{constant} \tag{2.16}$$

This constant is the capacitance given in farads, i.e.

$$C = \frac{Q}{V} \qquad (2.17)$$

Generally microfarads (10^{-6}), nanofarads (10^{-9}) or picofarads (10^{-12}) are used as the farad is a large unit.

There are occasions when capacitors have to be connected in series/parallel arrangements in order to achieve a particular capacitor value which is not a recognized manufacturer's value.

2.5 Capacitors in series

It is informative to explain the charge mechanism in Fig. 2.6. Plate A can acquire a positive charge by losing electrons to the supply, and the flux due to the charge on plate A terminates on plate B, which must therefore acquire a negative charge numerically equal to the positive charge on A. Plates B and C were previously neutral so now plate C must acquire a positive charge equal to the negative charge on D, which in turn causes a positive charge on plate E. Finally, this causes F to acquire a negative charge from the supply. The net effect is that each capacitor undergoes the same change in charge and since each initially had no charge, each capacitor acquires the same charge Q.

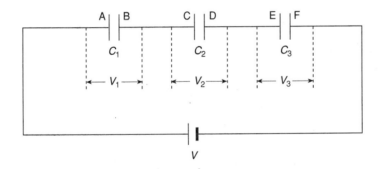

Figure 2.6

The sum of the p.d.s across the capacitors must satisfy Kirchhoff's voltage law:

$$V = V_1 + V_2 + V_3$$

$$\frac{Q}{C} = \frac{Q}{C_1} + \frac{Q}{C_2} + \frac{Q}{C_3} \qquad (2.18)$$

$$\therefore \frac{1}{C} = \frac{1}{C_1} + \frac{1}{C_2} + \frac{1}{C_3}$$

Note this is different from resistors in series.

2.6 Capacitors in parallel

In this case (Fig. 2.7), if a p.d. of V volts is applied to this parallel arrangement then a different charge appears on each capacitor:

$$\therefore Q = Q_1 + Q_2 + Q_3 = VC_1 + VC_2 + VC_3$$

$$\therefore \frac{Q}{V} = C = C_1 + C_2 + C_3 \qquad (2.19)$$

Thus the effective capacitance of the system is found as for resistors in series.

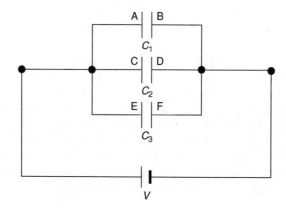

Figure 2.7

Example 2.2

Calculate the capacitance between points A and B in Fig. 2.8.

Solution

$$3000\,\text{pF} = 3\,\text{nF}$$

$$\therefore 3\,\text{nF} + 10\,\text{nF} = 13\,\text{nF} \quad \text{(as they are in parallel)}$$

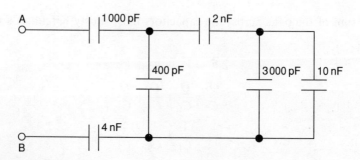

Figure 2.8

then

$$\frac{1}{13} + \frac{1}{2} = \frac{2+13}{26} = 1.7 \, \text{nF} \quad \text{(as they are in series)}$$

$$400 \, \text{pF} = 0.4 \, \text{nF}$$

$$0.4 + 1.7 = 2.1 \, \text{nF} \quad \text{(as they are in parallel)}$$

$$1000 \, \text{pF} = 1 \, \text{nF}$$

Therefore, total capacitance is

$$\frac{1}{1} + \frac{1}{2.1} + \frac{1}{4} = \frac{8.4 + 4 + 2.1}{8.4} = \frac{14.5}{8.4}$$

$$\therefore \ C = 0.58 \, \text{nF}$$

Example 2.3

A coaxial cable is run into a terminating cabinet alongside a single-core cable which does not have a charge. If the screen of the coaxial cable is not earthed, charges are induced as shown in Fig. 2.9. When a charge is applied resultant capacitances are measured and found to be

$$C_{AB} = 10 \, \text{pF} \qquad C_{BC} = 1.5 \, \text{pF}$$

$$C_{BE} = 5 \, \text{pF} \qquad C_{CE} = 2.2 \, \text{pF}$$

Determine the effective capacitance (C_E) in this system.

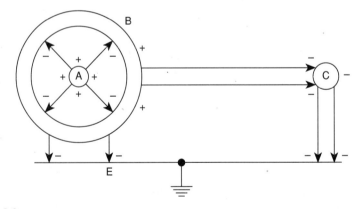

Figure 2.9

Solution
The system could be represented as shown in Fig. 2.10. Since $Q = CV$,

$$Q = C_1 V_1 = C_2 V_2$$

$$\frac{V_1}{V_2} = \frac{C_2}{C_1}$$

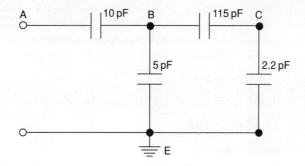

Figure 2.10

For the initial conditions

$$\frac{V_1}{140 - V_1} = \frac{C_2}{C_1}$$

$$\frac{70}{140 - 70} = \frac{C_2}{C_1} \qquad (2.20)$$

$$\frac{70}{70} = \frac{C_2}{C_1}$$

For the final conditions

$$\frac{75}{140 - 75} = \frac{C_2 + 4}{C_1}$$

$$75C_1 = 65(C_2 + 4) \qquad (2.21)$$

$$75C_1 = 65C_2 + 260$$

From (2.20)

$$C_2 = C_1$$

Substitute in (2.21)

$$75C_1 = 65C_1 + 260$$

$$\therefore \ C_1 = \frac{260}{10} = 26 \,\mu\text{F} = C_2$$

hence

$$\frac{1}{C_{CE}} + \frac{1}{C_{BC}} = \frac{1}{2.2} + \frac{1}{1.5}$$

$$= \frac{1.5 + 2.2}{3.3} = \frac{3.7}{3.3} = 1.12 \,\text{pF}$$

$$1.12 + 5 = 6.12 \,\text{pF}$$

Finally

$$\frac{1}{6.12} + \frac{1}{10} = \frac{10 + 6.12}{61.2}$$

$$\therefore \ C_E = 3.8\,\text{pF}$$

Example 2.4

Two capacitors C_1 and C_2 are connected in series across a 140 V d.c. switching circuit (Fig. 2.11). The p.d. across C_1 is 70 V, but this rises to 75 V when a $4\,\mu\text{F}$ capacitor is connected across C_2. Calculate the values of C_1 and C_2.

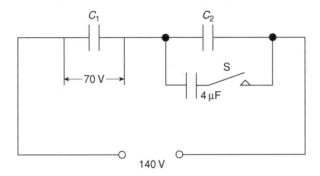

Figure 2.11

Solution

When capacitor is switched in 65 V appears across the $4\,\mu\text{F}$ capacitor.

$$\therefore \ Q = CV = \frac{4 \times 65}{10^6} = 260\,\mu\text{C}$$

The total charge in the parallel arrangement is the same as the charge on C_1 (capacitors in series)

$$\therefore \ 260 + 65C_2 = 75C_1$$

As 70 V appears across C_1 before switch S is closed then $C_1 = C_2$

$$\therefore \ 260 + 65C_1 = 75C_1$$

$$\therefore \ C_1 = C_2 = 26\,\mu\text{F}$$

2.7 *RC* time constants and d.c. transient response

When a voltage is applied across the plates of a capacitor it takes a finite time for this voltage to build up and finally equal the supply voltage. No current then flows externally as there is no p.d. in the circuit. The rate at which the capacitor voltage is growing can be compared to a voltage velocity and is written in its basic form as

$$\frac{\mathrm{d}V}{\mathrm{d}t} = \frac{I}{C} \tag{2.22}$$

Consider the RC circuit in Fig. 2.12. When the switch is closed there is no voltage across the capacitor and so current initially is

$$I_\mathrm{s} = \frac{V_\mathrm{s}}{R}$$

As the voltage (V_c) increases across the capacitor the current falls and

$$I = (V_\mathrm{s} - V_\mathrm{c})/R \tag{2.23}$$

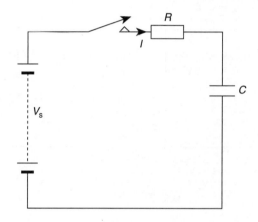

Figure 2.12

From equation (2.22)

$$I = C\,\mathrm{d}V/\mathrm{d}t$$

Substitute in (2.23)

$$C\,\mathrm{d}V/\mathrm{d}t = (V_\mathrm{s} - V_\mathrm{c})/R$$
$$\therefore\ \ \mathrm{d}V/\mathrm{d}t = (V_\mathrm{s} - V_\mathrm{c})/CR \tag{2.24}$$

The solution to this equation can be found in a standard mathematics text and it is given as

$$V = V_\mathrm{s}[1 - \mathrm{e}^{-t/CR}] \tag{2.25}$$

When $t = 0$, $\mathrm{e}^{-0} = 1$ so that $V_\mathrm{c} = 0$; when t becomes very large, $\mathrm{e}^{-t/CR}$ becomes very small so that V_c is nearly but never quite V_s. These points are shown in Fig. 2.13. Figure 2.13 shows what would have happened if the charge had continued to flow at the initial rate, i.e. the capacitor would have charged linearly as represented by OA to the supply voltage in a time T which is the time constant and is given as

$$T = RC \tag{2.26}$$

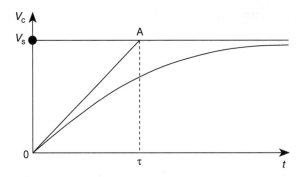

Figure 2.13

The following points can be determined from this exercise:

1. Increasing C has no effect on the initial charge or discharge current, while increasing R reduces the value of the initial charge or discharge current.
2. For most practical purposes the capacitor is assumed to reach its final voltage about 5τ.
3. The complete set of charge/discharge equations are given as

$$V_c = V_s(1 - e^{-t/\tau}) \quad \text{(charge)} \tag{2.27}$$

$$V_c = V_s\,e^{-t/\tau} \quad \text{(discharge)} \tag{2.28}$$

$$I = I_s\,e^{-t/\tau} \quad \text{(charge/discharge)} \tag{2.29}$$

Example 2.5

An RC network has to be switched into a 140 V d.c. circuit in order to decrease the initial rate of rise in voltage to 100 V/s. If the d.c. circuit has an equivalent resistance of 2.3 kW at its terminals and the ultimate charge across this capacitor has to be 0.035 C, determine:

(a) the value of C
(b) the value of R
(c) the initial charging current
(d) the charging current and voltage after 50 ms.

Solution
(a) Since $Q = CV$ then

$$C = \frac{Q}{V} = \frac{0.035 \times 10^6}{140} = 250\,\mu\text{F}$$

(b) The initial rate of rise in voltage at t is given by equation (2.24):

$$\frac{dV}{dt} = (V_s - V_c)/CR_T$$

where R_T is the total resistance in the network after switching, and as $V_c = 0$ at $t = 0$ then

$$\frac{\mathrm{d}V}{\mathrm{d}t} = \frac{V_s}{CR_T}$$

$$\therefore \ R = \frac{V_s}{C\,\mathrm{d}V/\mathrm{d}t}$$

$$= \frac{140}{250 \times 10^{-6} \times 100}$$

$$= \frac{14 \times 10^7}{25 \times 10^3}$$

$$= \frac{14 \times 10^4}{25} = 5.6\,\mathrm{k\Omega}$$

Since $R_T = R + 2.3$ then

$$R = R_T - 2.3$$

$$\therefore \ R = 3.3\,\mathrm{k\Omega}$$

(c) Initial charging current is given by equation (2.29):

$$\therefore \ I = I_s\,\mathrm{e}^{-t/CR} = \frac{V_s}{R_T}\,\mathrm{e}^{-t/CR_T}$$

Therefore, when $t = 0$,

$$I = \frac{V_s}{R_T} = \frac{140}{5.6 \times 10^3} = 25\,\mathrm{mA}$$

(d) $V_c = V_s(1 - \mathrm{e}^{-t/\tau})$

$$\frac{t}{\tau} = 3.3 \times 10^3 \times 10^3 \times 250 = 0.06$$

$$\therefore \ V_c = 140(1 - \mathrm{e}^{-0.06}) = 8.26\,\mathrm{V}$$

$$I = I_s\,\mathrm{e}^{-t/\tau} = 25\,\mathrm{e}^{-0.06} = 23.5\,\mathrm{mA}$$

Example 2.6

An attenuation network is placed between a 14 V supply and a switching load as shown in Fig. 2.14. The switch has to be moved to position 2 when $t = 100\,\mathrm{ms}$. Initially the switch is in position 1 at $t = 0$. Determine the voltage across the load when $t = 120\,\mathrm{ms}$.

Solution
Using Thévenin's theorem gives the equivalent resistance, R_T, as

$$\frac{10 \times 5}{10 + 5} = 3.3\,\mathrm{k\Omega} \quad \text{then} \quad 3.3 + 10 = 13.3\,\mathrm{k\Omega}$$

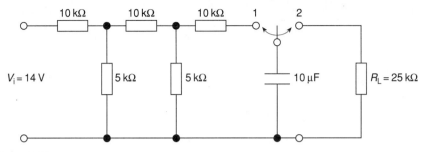

Figure 2.14

Finally

$$\frac{13.3 \times 5}{13.3 + 5} = 3.6\,\text{k}\Omega$$

The Thévenin generator V_T is determined from Figs 2.15 and 2.16.
  The original network now becomes that shown in Fig. 2.17. During charging

$$V_c = V(1 - e^{-t/CR} = 1.27(1 - e^{-t/\tau})$$

$$T = CR = 10 \times 10^{-6} \times 13.6 \times 10^{3} = 136\,\text{ms}$$

Hence

$$V_c = 1.27(1 - e^{-100/136}) = 0.66\,\text{V}$$

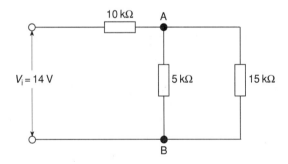

Figure 2.15

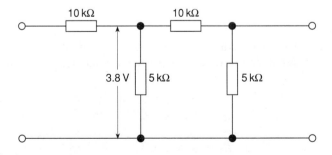

Figure 2.16

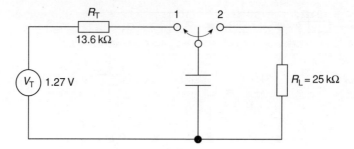

Figure 2.17

When the switch is moved to position 2 and the capacitor discharges through the load,

$$T = CR_L = 10 \times 10^{-6} \times 25 \times 10^3 = 250\,\text{ms}$$

When $t = 120\,\text{ms}$ this is 20 ms after the load has been switched:

$$\therefore \ V_c = 0.66\,e^{-t/\tau} = 0.66\,e^{-20/250} = 0.609\,\text{V}$$

Hence voltage across the load is $0.66 - 0.609 = 51\,\text{mV}$.

Example 2.7

The circuit in Fig. 2.18 has to supply a minimum output voltage of 150 mV 60 μs after switching on. Show that the network will provide this voltage.

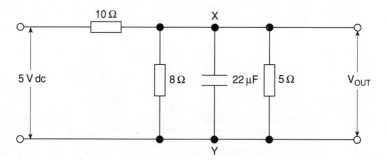

Figure 2.18

Solution

This network has to be configured as an *RC* transient network, hence the equivalent Thévenin circuit has to be derived across it. The Thévenin resistance is found from Fig. 2.19:

$$\frac{1}{R_T} = \frac{1}{10} + \frac{1}{8} + \frac{1}{5} = \frac{4+5+8}{40}$$

$$\therefore \ R_T = \frac{40}{17} = 2.35\,\Omega$$

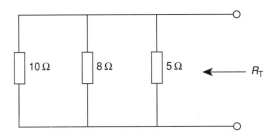

Figure 2.19

The Thévenin voltage is deduced from Fig. 2.20. This gives the familiar transient circuit as shown in Fig. 2.21:

$$V_T = \left[\frac{8 \times 5}{8 + 5} \middle/ \left(\frac{8 \times 5}{8 + 5} + 10\right)\right] \times 5$$

$$= 0.235 \, \text{V}$$

$$\therefore V_{out} = 0.235(1 - \exp[6 \times 10 \times 10^6 / 2.35 \times 2.21 \times 10^6])$$

$$= 0.235(1 - e^{-1.16}) = 0.235(1 - 6.313)$$

$$= 0.16 = 160 \, \text{mV}$$

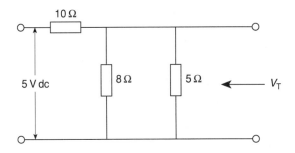

Figure 2.20

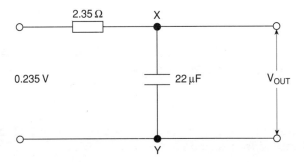

Figure 2.21

2.8 AC transient response

Section 2.7 examined the effects of d.c. transients on an RC network. We now wish to consider the effect of a time-varying voltage applied to such a network, as there are many applications where sinusoidal, square and step pulses are present as well as d.c. transients. Examples of this are RC compensation networks for amplifiers, motor sequence control, integrator networks, differential networks, power supply filtering, power factor correction, decoupling and modulators.

2.9 Capacitive reactance

Before looking at some of the various applications mentioned above the concept of capacitive reactance should be understood. If a voltage is applied to a capacitor it can be seen from (2.22) that the instantaneous current value is given by

$$i = C \frac{dV}{dt}$$

and since the voltages now being considered are sinusoidal

$$i = C \frac{d}{dt} (V_m \sin 2\pi ft)$$

$$= 2\pi f C V_m \cos 2\pi ft$$

$$= 2\pi f C V_m \sin \left(2\pi ft + \frac{\pi}{2} \right) \qquad (2.30)$$

Expression (2.30) shows two points, i.e. the current leads the voltage by 90° and also the maximum current is given by

$$I_m = V_m 2\pi f C$$

If I_m and V_m are both r.m.s. values then

$$\frac{V}{I} = \frac{1}{2\pi f C} = X_c = \text{capacitive reactance} \qquad (2.31)$$

2.10 Growth of current in an *RC* network with an applied sinusoidal voltage

Section 2.7 considered the concept of d.c. transients, but as has been mentioned in section 2.8 most transient circuits contain d.c. and a.c. elements. A full analysis is required for this more complex question. Figure 2.22 shows an RC series circuit excited from a sinusoidal source (V) where

$$v = V_m \sin(wt + \theta) \qquad (2.32)$$

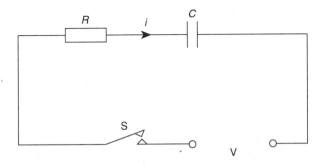

Figure 2.22

where θ is referred to as the switching angle which gives the value of the instantaneous voltage at the instant the switch is closed, i.e. $t = 0$. Thus at $t = 0$

$$v = V_m \sin \theta \tag{2.33}$$

or

$$\theta = \sin^{-1} \left(\frac{v}{V_m} \right) \tag{2.34}$$

As the instantaneous value of current is a combination of a steady-state sinusoidal current (i_{ss}) and a transient component (i_t), then

$$i = i_{ss} + i_t = i_{ss} + A\,e^{-t/RC} \tag{2.35}$$

Note that A is not simply I_s as given in expression (2.29). This constant has to be derived. Now $i_{ss} = v/Z$ where

$$Z = \sqrt{R^2 + X_c^2} \angle - \phi$$

$$\phi = \tan^{-1} \left(\frac{X_c}{R} \right) \quad \text{(phase angle)}$$

$$\therefore \; i_{ss} = \frac{v}{Z} = V_m \sin(wt + \theta)/Z \angle - \phi$$

$$= \frac{V_m}{Z} \sin(wt + \theta + \phi)$$

$$\therefore \; i_{ss} = I_m \sin(wt + \theta + \phi) \tag{2.36}$$

where $I_m = V_m/Z$. Thus

$$i = I_m \sin(wt + \theta + \phi) + A\,e^{-t/CR} \tag{2.37}$$

In order to evaluate A consider the situation at $t = 0$. At this moment all the supply voltage is across the resistor.

$$\therefore \; i = \frac{v}{R} = \frac{V_m \sin \theta}{R}$$

$$\frac{V_m \sin \theta}{R} = I_m \sin(\theta + \phi) + A\,e^0$$

Thus at $t = 0$

$$A = \frac{V_m \sin \theta}{R} - \frac{V_m}{Z}(\sin \theta \cos \phi + \cos \theta \sin \phi)$$

$$A = \frac{V_m}{R}\left[\sin \theta - \frac{R}{Z}(\sin \theta \cos \phi + \cos \theta \sin \phi)\right]$$

since $R/Z = \cos \phi$, then

$$A = \frac{V_m}{R}[\sin \theta - \cos \phi(\sin \theta \cos \phi + \cos \theta \sin \phi)]$$

$$= \frac{V_m}{R}[\sin \theta - \sin \theta \cos^2 \phi - \cos \theta \sin \phi \cos \phi]$$

$$= \frac{V_m}{R}[\sin \theta(1 - \cos^2 \phi) - \cos \theta \sin \phi \cos \phi]$$

Since $\sin^2 \phi + \cos^2 \phi = 1$ then $1 - \cos^2 \phi = \sin^2 \phi$:

$$A = \frac{V_m}{R}[\sin \theta \sin^2 \phi - \cos \theta \sin \phi \cos \phi]$$

$$= \frac{V_m}{R}[\sin \phi(\sin \theta \sin \phi - \cos \theta \cos \phi)]$$

$$= \frac{-V_m}{R}[\sin \phi(\cos \theta \cos \phi - \sin \theta \sin \phi)]$$

$$= \frac{-V_m}{R}\sin \phi \cos(\theta + \phi)$$

Substituting for A in (2.35) gives

$$i_t = \frac{V_m}{R}\sin \phi \cos(\theta + \phi)\, e^{-t/CR} \tag{2.38}$$

From (2.36) and (2.38) the total current is

$$i = I_m \sin(wt + \theta + \phi) - \frac{V_m}{R}\sin \phi \cos(\theta + \phi)\, e^{-t/CR} \tag{2.39}$$

From expression (2.39) note the following effects the switching angle has on the magnitude of the transient:

1. If $\theta = -\phi$ then $\cos(\theta + \phi) = \cos(-\phi + \phi) = \cos 0$. Thus the transient will be a maximum negative.
2. If $\theta = -\phi + 180°$ then $\cos(\theta + \phi) = \cos(-\phi + 180) = \cos(\pi)$. Thus the transient will be a maximum positive.
3. If $\theta = 90 - \phi$ then $\cos(\theta + \phi) = \cos(90 - \phi + \phi) = \cos(90°) = 0$. Thus the transient current will be zero.

Example 2.8

In the circuit shown in Fig. 2.23 the switch is closed when the voltage is 141.5 V and going positive.

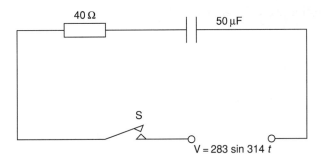

Figure 2.23

(a) Determine an expression for the current at a time t seconds after the switch closes.

(b) Calculate the current at $t = 10\,\mathrm{ms}$ after the switch closes.

Solution

(a) From (2.39)

$$i = I_m \sin(wt + \theta + \phi) - \frac{V_m}{R} \sin \phi \cos(\theta + \phi)\, e^{-t/CR}$$

$$X_c = \frac{1}{2\pi fc} = \frac{1}{314 \times 5 \times 10^{-6}} = 63.7\,\Omega$$

$$Z = \sqrt{40^2 + 63.7^2} = 75.2\,\Omega \qquad \theta = \tan^{-1}\frac{X_c}{R} = \frac{63.7}{40}$$

$$I_m = \frac{V_m}{Z} = \frac{283}{75.2} = 3.8\,\mathrm{A}$$

$$\frac{V_m}{R} = \frac{283}{40} = 7.08\,\mathrm{A}$$

$$\theta = \sin^{-1}\frac{v}{V_m} \text{ at } t = 0, \text{ i.e. } \theta = \sin^{-1}\left(\frac{141.5}{283}\right)$$

$$\theta + \phi = 0.52 + 1 = 1.52\,\mathrm{rad}$$

$$\cos(\theta + \phi) = \cos(1.52) = 0.05$$

$$\sin \phi = \sin(-1) = 0.84$$

$$CR = 40 \times 50 \times 10^{-6} = 0.002\,\mathrm{s}$$

$$\frac{1}{CR} = 500 \text{ per second}$$

Substituting these values in (2.39) gives

$$i = I_m \sin(wt + \theta + \phi) - \frac{V_m}{R} \sin\phi \cos(\theta + \phi)\, e^{-t/CR}$$

$$= 3.8\sin(314t + 0.52 + 1) - 7.08\sin 58° \cos(29.7 + 58)\, e^{-t/500}$$

$$= 3.8\sin(314t + 1.52) - 7.08\sin(1)\cos(1.52)\, e^{-t/500}$$

(b)
$$i = 3.8\sin\left(\frac{314 \times 10}{10^3} + 1.52\right) - 7.08\sin(1)\cos(1.52)\, e^{-10/5 \times 10^5}$$

$$= 3.8\sin 4.66 - 7.08\sin(1)\cos(1.52)\, e^{-2 \times 10^{-5}}$$

$$= -3.79 - 7.08 \times 0.84 \times 0.05\, e^{-2 \times 10^{-5}}$$

$$= -3.79 - 0.29$$

$$\therefore\ i = -4.08\,\text{A}$$

2.11 AC bridge networks

In Chapter 1 the Wheatstone bridge was examined and a few of its practical applications considered. AC bridges are a natural extension of this bridge. There are many types of bridge, but they all have four arms, impedance, a power supply and a detector. Figure 2.24 shows the general structure. The main difference from d.c. bridges is the inclusion of impedances which will consist, in this chapter, of RC networks.

As with the Wheatstone bridge, balance will occur when the detector shows zero current. The same current then passes through Z_4 and Z_3, while the

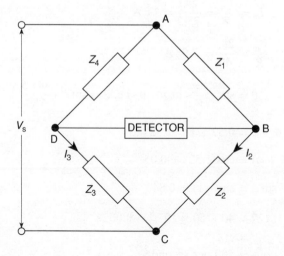

Figure 2.24

same current passes through Z_1 and Z_2. This gives

$$I_2Z_1 = I_3Z_4$$

$$I_2Z_2 = I_3Z_3$$

and dividing gives

$$\frac{I_2Z_1}{I_2Z_2} = \frac{I_3Z_4}{I_3Z_3} \tag{2.40}$$

$$\frac{Z_1}{Z_2} = \frac{Z_4}{Z_3} \tag{2.41}$$

$$Z_1Z_3 = Z_2Z_4$$

When deriving these balance equations for an a.c. bridge, where two of the arms at least are in complex form, a complex equation will be produced such as (2.42):

$$a + jb = c + jd \tag{2.42}$$

In such an equation the real parts are equal and also the imaginary parts. Hence

$$a = c \quad \text{and} \quad b = d$$

2.12 The method of deriving the balance equations for any a.c. bridge

It is not the purpose of this text to examine the large selection of bridges which are used in industrial testing and measurement. However, a few of the more frequently used ones will be considered.

As the object of all bridges is initially to achieve balance, the following procedure should be adopted:

1. Determine the impedance in each arm in complex and real form.
2. Form the complex equation using (2.42).
3. Equate the real and imaginary parts of the equation.
4. Finally, determine the components and operational frequency.

In practice one arm of the bridge contains the unknown impedance, while the other arms consist of known fixed or variable components. Generally the current in the detector is reduced to zero by adjustment of one or two variable components.

2.13 Practical *RC* bridges

Three common bridges which are used frequently are:

1. The de Sauty bridge
2. The Wien bridge
3. The Schering bridge.

The de Sauty bridge

This bridge (Fig. 2.25) is one of the simplest and is used to measure an unknown capacitor by comparing it with a known one (C_3). At balance $ZZ_2 = Z_1Z_3$, i.e.

$$(-jXc)R_2 = R_1(-jXc_3)$$

$$\therefore\ R_2Xc = Xc_3R_1$$

$$R_2\left(\frac{1}{wC}\right) = R_1\left(\frac{1}{wC_3}\right)$$

This gives

$$\frac{R_2}{C} = \frac{R_1}{C_3} \quad \text{or} \quad C = \frac{R_2C_3}{R_1} \tag{2.43}$$

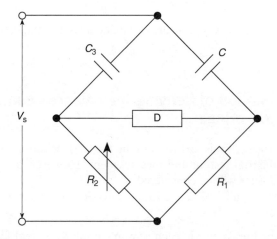

Figure 2.25

The Wien bridge

This bridge (Fig. 2.26) is used to measure frequency or capacitance depending on what is known. It is commonly used in oscillator circuits where stabilization is required. In this case,

$$Z_1 = R_1 \qquad Z_2 = \frac{-jXc_2 \times R_2}{-jXc_2 + R_2}$$

$$Z_4 = R_4$$

$$Z_3 = R_3 - jXc$$

$$= R_3 - \frac{j}{wC_3} = \frac{1}{(1/R_2) + jwC_2}$$

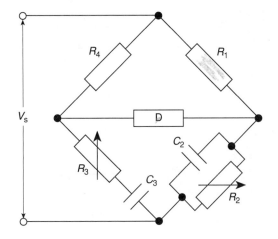

Figure 2.26

At balance $Z_1 Z_3 = Z_2 Z_4$

$$(R_1)(R_3 - jXc_3) = \left(\frac{1}{(1/R_2) + jwC_2} \right)(R_4)$$

Rearranging gives

$$\left(R_3 - \frac{1}{wC_3} \right) = \left(\frac{1}{R_2} + jwC_2 \right) = \frac{R_4}{R_1}$$

$$\therefore \quad \frac{R_3}{R_2} + \frac{C_2}{C_3} - j\left(\frac{1}{wC_3 R_2} \right) + jwC_2 R_3 = \frac{R_4}{R_1}$$

Equating real parts gives

$$\frac{R_3}{R_2} + \frac{C_2}{C_3} = \frac{R_4}{R_1} \tag{2.44}$$

Equating imaginary parts gives

$$-\frac{1}{wC_3 R_2} + jwC_2 R_3 = 0$$

Transposing gives

$$w^2 = \frac{1}{C_2 C_3 R_2 R_3}$$

$$\therefore \ f = \frac{1}{2\pi \sqrt{C_2 C_3 R_2 R_3}} \tag{2.45}$$

In many applications $C_2 = C_3$ and $R_2 = R_3$ so that (2.44) becomes

$$f = \frac{1}{2\pi CR} \tag{2.46}$$

The Schering bridge

This bridge (Fig. 2.27) is used to measure the capacitance of a capacitor and its equivalent series resistance. Unlike the other two bridges, it can also measure the power factor of insulating materials and dielectric losses.

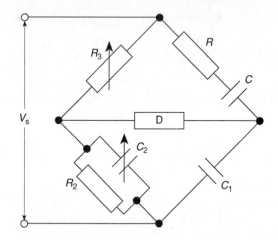

Figure 2.27

In Fig. 2.27, C is the unknown capacitor and R is its equivalent series resistance due to the dielectric material. At balance

$$Z = R - jXc \qquad Z_1 = -jXc_1$$

$$Z_2 = \frac{(R_2)(-jXc_2)}{(R_2 - jXc_2)} \qquad Z_3 = R_3$$

At balance

$$(Z)(Z_2) = (Z_1)(Z_3)$$

$$(R - jXc)\frac{(R_2)(-jXc_2)}{R_2 - jXc_2} = (-jXc_1)(R_3)$$

Transposing gives

$$(R - jXc) = \frac{Xc_1 R_3}{Xc_2 R_2}(R_2 - jXc_2)$$

Equating the real parts gives

$$R = \frac{Xc_1 R_3}{Xc_2} = \frac{(1/wC_1)R_3}{(1/wC_1)} = \frac{C_2 R_3}{C_1} \tag{2.47}$$

Equating imaginary parts gives

$$-Xc = \frac{-Xc_1 R_3}{R_2}$$

$$\frac{1}{wC} = \frac{(1/wC_1)R_3}{R_2} = \frac{R_3}{wC_1 R_2}$$

$$\therefore \ C = \frac{C_1 R_2}{R_3} \tag{2.48}$$

It can also be shown from a simple phasor diagram that

$$\phi = \arctan\left(\frac{1}{wCR}\right) \tag{2.49}$$

hence the power factor $\cos\phi$ can be found.

Example 2.9

A Wien bridge has the following value:

$$R_2 = R_3 = 56\,\text{k}\Omega \qquad C_2 = C_3 = 2500\,\text{pF} \quad \text{and} \quad R_4 = 1.2\,\text{k}\Omega$$

Determine under balance conditions the value of R_1 and the frequency of operation.

Solution
From equation (2.44)

$$\frac{R_3}{R_2} + \frac{C_2}{C_3} = \frac{R_4}{R_1}$$

$$1 + 1 = \frac{1200}{R_1}$$

$$\therefore \ R_1 = 600\,\Omega$$

Also

$$f = \frac{1}{2\pi RC} = \frac{10^{12}}{6.28 \times 56 \times 10^3 \times 2.5}$$

$$\therefore \ f = 11.376\,\text{kHz}$$

Example 2.10

A Schering bridge having a frequency of 3.2 kHz has to be used in order to find the following parameters of a capacitor:

(a) The value of its equivalent resistance
(b) The value of the capacitor
(c) The phase angle of the *CR* arm
(d) The power factor of the *CR* arm.

The configuration is shown in Fig. 2.28.

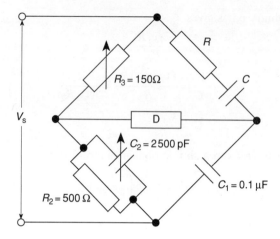

Figure 2.28

Solution

From equation (2.47):

(a)
$$R = \frac{C_2 R_3}{C_1} = \frac{25 \times 10^2 \times 150 \times 10^6}{0.1 \times 10^{12}} = 3.7\,\Omega$$

Also from (2.48)

(b)
$$C = \frac{C_1 R_2}{R_3} = \frac{0.1 \times 5 \times 10^2}{10^6 \times 150} = \frac{5}{15} = 0.33\,\mu F$$

From (2.49)

(c)
$$\phi = \arctan\left(\frac{1}{wCR}\right)$$

$$= \arctan \frac{10^6}{6.28 \times 3.2 \times 10^2 \times 0.33 \times 3.7}$$

$$= \arctan \frac{10^4}{24.87} = 89.85°$$

(d)
$$\therefore \quad \cos\phi = 0.00248$$

2.14 Power supply ripple filters

At the beginning of this chapter the question was posed regarding the filtering of the ripple associated with power supplies. A basic power supply is shown in block form in Fig. 2.29. This power supply is essentially an a.c. to d.c. converter, the d.c. load voltage being stabilized by the regulator. Ideally, the load functions with d.c., but there is an a.c. component superimposed on this called the ripple. The smaller this ripple the better the regulation of the supply. This is the function of the filter block.

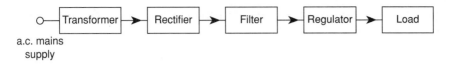

a.c. mains
supply

Figure 2.29

One of the basic forms of ripple filter is a large capacitor, generally electro-lytic, which is connected as shown in Fig. 2.1. The capacitor filter is cheap and simple, but it has the disadvantage that it takes current from the trans-former in a series of pulses.

In order to design the power supply for improved regulation and stability the filter capacitor has to be properly chosen. In calculating its value the ideal output voltage waveform is analysed in Fig. 2.30.

As we have seen already, a capacitor charges and discharges exponentially, but for design purposes the sudden charge and discharge lines shown give an accurate enough result. During the discharge period the output voltage falls at a reasonably constant rate, falling from the peak value (V_p) at the rate of

$$\frac{dV}{dt} = \frac{V_p}{R_L C}$$

where R_L is the load resistance and C the capacitance value; $R_L C$, remember, is the time constant (T).

As δV_L is the peak-to-peak ripple voltage superimposed on the d.c. or load voltage (V_L)

$$\delta V_L = \left(\frac{V_p}{R_L C}\right) T \tag{2.50}$$

The d.c. or load voltage is given approximately by

$$V_L - V_p = \frac{-\delta V_L}{2} = V_p\left(1 - \frac{T}{2R_L C}\right) \tag{2.51}$$

Equation (2.51) enables C to be determined for a half-wave rectifier. For a full-wave rectifier which is normally used in practical supplies, the output in Fig. 2.30

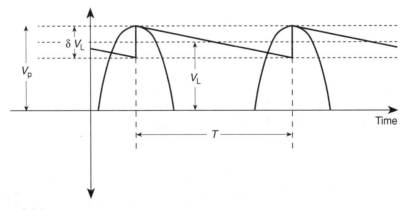

Figure 2.30

pulses at twice the frequency, hence the period of discharge is $T/2$ and equation (2.51) becomes

$$V_L = V_p \left(1 - \frac{T}{4R_L C} \right) \qquad (2.52)$$

A second method of calculating this capacitor is given from expressions (2.1) and (2.17). From the definition of current:

$$i = \frac{dq}{dt} = \frac{C \, dv}{dt} = I_p$$

In this case dv is the peak-to-peak ripple voltage, previously given by δVL, while dt is the time between pulses, namely T or $T/2$ depending on whether the supply is full or half-wave.

Once again the value of C can be calculated and this will give the same value as for method 1.

Example 2.11

A full-wave power supply has to be designed with the following parameters:

$$\text{Load resistance} = 115\,\Omega$$

$$\text{Load voltage} = 45\,\text{V}$$

$$\text{a.c. supply voltage} = 240\,\text{V}/50\,\text{Hz (r.m.s.)}$$

What value of filter capacitor would be suitable for this supply?

Solution
From equation (2.52)

$$V_L = V_p \left(1 - \frac{T}{4R_L C} \right)$$

Rearranging gives

$$C = \frac{T}{4R_L} \left(\frac{1}{1 - V_L/V_p} \right)$$

$$V_p = \sqrt{2} \times 240 = 339$$

$$T = -\frac{1}{f} = \frac{1}{50} = 0.02\,\text{s}$$

$$\therefore C = \frac{0.02}{4 \times 115} \left(\frac{1}{1 - 45/339} \right)$$

$$= \frac{0.02}{460} \left(\frac{1}{0.867} \right)$$

$$= \left(\frac{0.02}{460} \right) 1.153 = 50\,\mu\text{F}$$

Example 2.12

A full wave 320 V/50 Hz supply feeds a resistance load which draws a current of 5.2 A when the load voltage is 310 V. Determine the value of capacitor necessary to handle the ripple voltage involved.

Solution
From equation (2.51)

$$V_L = V_p - \frac{\delta V_L}{2}$$

Rearranging gives

$$\frac{\delta V_L}{2} = V_p - V_L = 320 - 310$$

$$\therefore \; \delta V_L = 20 \text{ V}$$

Since $R_L = V_L/I_L$ then

$$R_L = \frac{310}{5.2} = 59.6 \,\Omega$$

From equation (2.50)

$$\delta V_L = \left(\frac{V_p}{R_L C}\right) T$$

$$\therefore \; C = \left(\frac{V_p}{\delta V_L R_L}\right) T$$

$$= \left(\frac{320}{20 \times 59.6}\right) 0.01 = 26.845 \,\mu\text{F}$$

This application shows that large capacitors are required where high currents with small ripple factors are required.

2.15 Pulse response of the *RC* network

When a pulse is connected to the input of an *RC* configuration the output response depends on whether the network is used as an integrator or a differentiator. This has important applications in wave shaping and timing pulse circuits.

Figure 2.31(a) shows a differentiator circuit while Fig. 2.31(b) is an integrator circuit. These circuits generally form part of an amplifier circuit which will be considered in Chapter 3. Figure 2.32 shows the waveform changes which occur when a 1 kHz rectangular wave is applied to the input of Fig. 2.31(a) for three different time constants. When a 1 kHz rectangular waveform is applied to Fig. 2.31(b) the same three time constants are taken to show the effect on the input waveform. As can be seen from Fig. 2.33, the differentiator output response depends on the time constant.

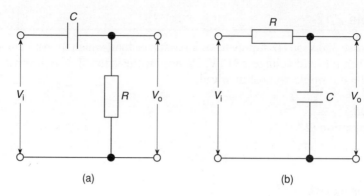

Figure 2.31

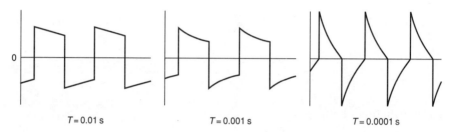

Figure 2.32

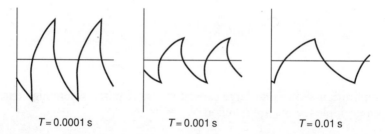

Figure 2.33

2.16 The *RC* integrator

Two conditions of the pulse response must be considered when considering this network:

1. When the input pulse width is equal to or greater than five time constants.
2. When the input pulse width is less than five time constants.

Note that five time constants are accepted as the time for a capacitor to fully charge or fully discharge and is called the transient time.

Figure 2.34 shows the two conditions mentioned above when an integrator network is used. Note in Fig. 2.34(a) that the output pulse rises to 5 V in each case; but the shorter the time, i.e. 2 μs, the more the shape of the output pulse

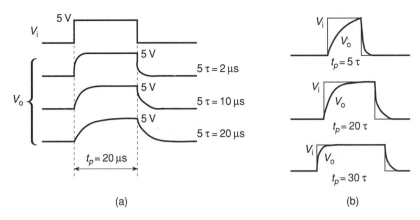

Figure 2.34

approaches the input. Figure 2.34(b) shows a fixed time constant with a variable input pulse width (t_p). As the width of the pulse is increased the shape of the output pulse approaches it.

If the second case is considered where the pulse width is less than five time constants then the capacitor will charge and discharge at different rates depending on the value of the time constant. This is shown in Fig. 2.35(a). Figure 2.35(b) illustrates what happens when the input pulse width for a fixed time constant is varied: the output voltage is reduced the narrower the pulse.

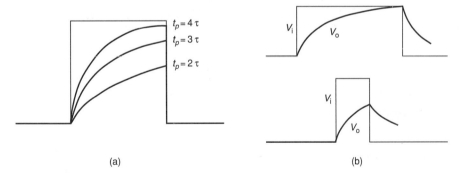

Figure 2.35

Example 2.13

A pulse generator produces a 5 V pulse having a duration of 100 µs which is applied to an integrator having $R = 10\,\text{k}\Omega$ and $C = 0.01\,\mu\text{F}$.

(a) Determine the voltage to which the capacitor will charge.
(b) Determine the time it takes for the capacitor to discharge through the source if its internal resistance is 600 Ω.
(c) Sketch the output voltage.

Solution

(a) The time constant is

$$\tau = 10^4 \times 0.01 \times 10^{-6} = 100\,\mu s$$

Since the pulse width is equal to one time constant the capacitor will charge to 63 per cent of its input amplitude in this time:

$$V_o = 0.63 \times 5 = 3.15\,V$$

(b) The capacitor discharges back through the source:

$$\therefore\ 5\tau = 5C(10^4 + 600)$$

$$= 5 \times 0.01 \times 10^6 \times 10^2 \times 10^{-6} = 530\,\mu s$$

(c) The output waveform is shown in Fig. 2.36.

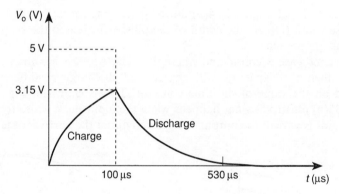

Figure 2.36

Example 2.14

The maximum possible output level which may be applied to a pulse-shaping circuit as shown in Fig. 2.37 is 9.5 V. Determine the values of R and C which will satisfy this condition.

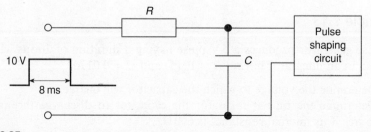

Figure 2.37

Solution

Select a value of R at random, say $3.6\,\text{k}\Omega$. Since

$$v = V(1 - e^{-t/CR})$$

$$\frac{v}{V} = 1 - e^{-t/CR}$$

$$\left(1 - \frac{v}{V}\right) = e^{-t/CR}$$

$$\ln\left(1 - \frac{v}{V}\right) = -t/CR$$

$$CR = \frac{-t}{\ln(1 - v/V)}$$

$$\therefore\ C = \frac{-t}{R\ln(1 - v/V)}$$

$$= \frac{8}{10^3 \times 3.6 \times 10^3 \ln(1 - 9.5/10)}$$

$$= \frac{8}{10^3 \times 3.6 \times 10^3 \times 0.693} = 3.2\,\mu\text{F}$$

2.17 The *RC* differentiator

The differentiator circuit shown in Fig. 2.31(a) produces waveforms similar to those found in Fig. 2.32. Once again two conditions of the pulse response are considered, namely:

1. When $t_p \geq 5\tau$.
2. When $t_p < 5\tau$.

The effects of this are shown in Fig. 2.38. In each case the response to the rising edge, between the rising and falling edges and the falling pulse edge should be noticed.

Example 2.15

A differentiator network consists of an *RC* network connected to a pulse-shaping circuit having an input resistance of $750\,\Omega$ (Fig. 2.39). If the input pulse has a height of $10\,\text{V}$ and a duration of $1\,\text{ms}$, determine the voltage waveform which will be applied to the input of the pulse-shaping circuit.

Solution

$$\tau = \frac{(2.2 \times 0.75)}{(2.2 + 0.75)} \times \frac{1}{10^6} \times 10^3 = 0.56\,\text{ms}$$

During the rising edge of the input pulse the voltage across the equivalent resistance of $560\,\Omega$ immediately rises to $10\,\text{V}$. The pulse width is $1\,\text{ms}$ so the

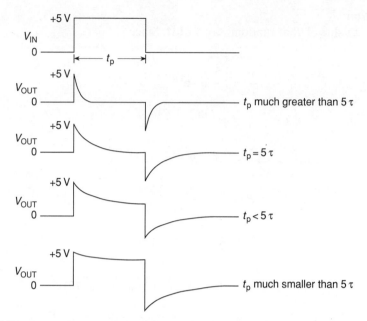

Figure 2.38

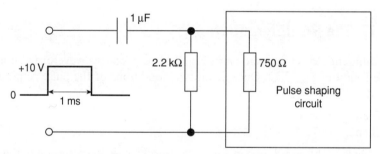

Figure 2.39

capacitor charges for 1.79 time constants approximately and so does not reach the full charge voltage. Hence at the end of the 5 ms pulse the output voltage is

$$V_{\text{out}} = V_{\text{in}}\, e^{-t/RC} = 10\, e^{-1/0.56} = 1.68\,\text{V}$$

On the falling edge the resistor voltage immediately jumps from +1.68 V down to 8.32 V (to a 10 V transition). The resulting waveform is shown in Fig. 2.40.

2.18 Capacitance sensors

In section 2.13 a method of processing changes in capacitance was examined by using *RC* bridges. Commercially produced capacitors were assumed to be used in these bridges, but capacitive sensors may be used as part of the bridge configuration. One such sensor is shown in Fig. 2.41.

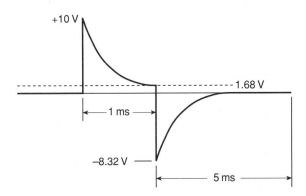

Figure 2.40

A physical capacitor depends on the distance between the plates (d), the area of the plates (A) and the dielectric constant (ε_r) of the non-conductive material between the plates. This is generally expressed mathematically as

$$C = \frac{\varepsilon_0 \varepsilon_r A}{d}$$

where ε_0 = the permittivity of free space, 8.854×10^{-12} F.

All these parameters are used in capacitive sensors and the one shown in Fig. 2.41 uses the variation of plate area (A) and the dielectric constant (ε_r) in order to sense linear or rotary movement. As the object moves, the dielectric sheet attached to the moving object slides between the plates and this varies the capacitance. Alternatively, the moving object can be coupled to one of the capacitor's metal plates so that any object movement causes a change in overlap and hence area.

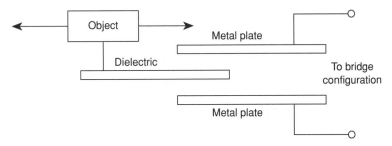

Figure 2.41

A second type of capacitance sensor utilizes a variation in plate separation and this can be used in a particular type of pressure sensor. One type is shown in Fig. 2.42. In this sensor the measured pressure is compared with the reference or atmospheric pressure. The resultant pressure causes the metal diaphragm to be compressed, thus causing the top plate to move downwards. The capacitance will increase as the distance has decreased. Conversely, a pressure less than atmospheric will cause the distance between the plates to increase, hence the capacitor decreases.

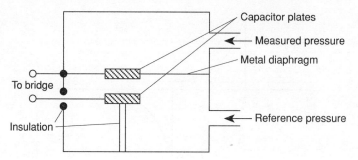

Figure 2.42

Example 2.16

A capacitive displacement sensor is shown in Fig. 2.43. If the plates are 50 mm square and the mica dielectric has to be capable of moving a distance of 40 mm, determine the changes in capacitance which a bridge must be capable of handling (ε_r for mica is 5).

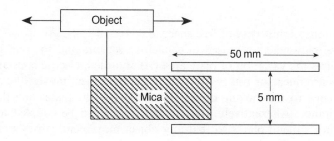

Figure 2.43

Solution

When the dielectric moves the sensor is equivalent to having two capacitors in parallel, as shown in Fig. 2.44. The capacitance before the dielectric moves is

$$C = \frac{\varepsilon_0 \varepsilon_r A}{d}$$

$$= \frac{8.854 \times 10^{-12} \times 5 \times 25 \times 10^2 \times 10^3}{10^6 \times 5}$$

$$= 22.135 \, \text{pF}$$

At maximum travel the capacitance will be as in Fig. 2.44(b):

$$C_M = \frac{\varepsilon_0 \varepsilon_r A}{d}$$

$$= \frac{8.854 \times 10^{-12} \times 5 \times 10^2 \times 10^3}{10^6}$$

$$= 0.8854 \, \text{pF}$$

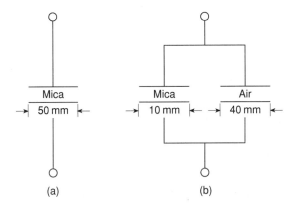

Figure 2.44

$$C_A = \frac{\varepsilon_0 A}{d}$$

$$= \frac{8.854 \times 10^{-12} \times 16 \times 10^2 \times 10^3}{10^6 \times 5}$$

$$= 28.33\,\text{pF}$$

As the air and mica capacitor are in parallel

$$C_A + C_M = 28.33 + 0.8854$$

$$= 29.218\,\text{pF}$$

Hence the bridge will see a change from 22.135 to 29.218 pF.

2.19 Further problems

1. The deflection plates of an ink-jet printer are set horizontally and parallel. The plates are square of side 1.5 cm and the field between them is uniform and of magnitude $1.5 \times 10^6\,\text{V}\,\text{m}^{-1}$. The mass of an ink drop is $1.2 \times 10^{-10}\,\text{kg}$ and has a charge $-1.6 \times 10^{-13}\,\text{C}$. The velocity of the ink drop is $20\,\text{m}\,\text{s}^{-1}$ on entering the plates. Determine the vertical displacement of the drop when it has completed its passage between the plates.
   Answer: 56 mm

2. An ink drop with a mass of $1.3 \times 10^{-10}\,\text{kg}$ and a negative charge of magnitude $1.5 \times 10^{-13}\,\text{C}$ enters the region between the plates, initially moving along the x-axis with a speed of $18\,\text{m}\,\text{s}^{-1}$. The length L of the plates is 1.6 cm and the electric field strength has a value of $1.4 \times 10^6\,\text{V}\,\text{m}^{-1}$. What will be the vertical deflection of the drop at the edge of the plates?
   Answer: 64 mm

3. Find the resultant capacitance between A and B in Fig. 2.45.
   Answer: $10\,\mu\text{F}$

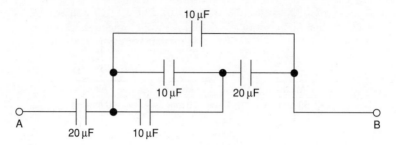

Figure 2.45

4. Find the total capacitance between the terminals in Fig. 2.46.
   Answer: $2\,\mu F$

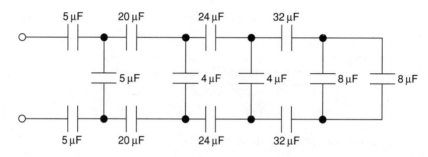

Figure 2.46

5. A capacitor of $100\,\mu F$ is connected in series with an $8\,k\Omega$ resistance. If the combination is connected suddenly to a $100\,V$ d.c. supply, find:
   (a) the initial rate of rise of p.d. across the capacitor;
   (b) the initial charging current;
   (c) the ultimate charge on the capacitor.
   Answer: $125\,V\,s^{-1}$, $12.5\,mA$, $0.01\,C$

6. A $2\,\mu F$ capacitor is joined in series with a $2\,M\Omega$ resistor to a d.c. supply of $100\,V$. Draw a current–time graph and calculate the current flowing in the capacitor after $t = 4\,s$ from the start.
   Answer: $18.4\,\mu A$

7. A $20\,\mu F$ capacitor is found to have an insulation resistance of $50\,M\Omega$, measured between the terminals. If this capacitor is connected to a $230\,V$ supply find, after disconnection, the time for the p.d. across the capacitor to fall to $60\,V$.
   Answer: $1342\,s$

8. A series circuit comprises a $1000\,\Omega$ resistor and a $5\,\mu F$ capacitor which is connected across a $10\,V$ rectangular pulse of width $3\,ms$. Determine the capacitor voltage and the circuit current $6\,ms$ after the rising edge of the pulse.
   Answer: $2.47\,V$, $2.47\,mA$

9. A series circuit comprises a 10 kΩ resistor and a 1 µF capacitor. A sinusoidal voltage given by $V = 100 \sin(1000t + \theta)$ is connected across the circuit at $t = 0$. Obtain an expression for the instantaneous current in the circuit for $t > 0$. If $v = 40$ V at $t = 0$, determine:
   (a) the current at $t = 4$ ms;
   (b) the voltage across the capacitor at $t = 4$ ms.
   Answer: −9.4 mA, −94 V

10. A 10 µF capacitor and a 10 kΩ resistor are connected in series across a 500 V d.c. supply as part of a switching circuit. The CR circuit is disconnected when the capacitor is fully charged and then switched into a 1 kΩ resistor circuit. Determine:
   (a) the initial value of the charging current in the circuit;
   (b) the initial value of the discharge current in the circuit;
   (c) the time taken for the voltage across the discharging capacitor to reach one-half of its initial value.
   Answer: 50 mA, 45.5 mA, 76.2 ms

11. In a Schering bridge network $C_1 = 0.2$ µF, $R_3 = 200$ Ω, $R_2 = 600$ Ω and $C_2 = 4000$ pF. If the supply frequency is 1.5 kHz, determine when the bridge is balanced:
   (a) the value of R;
   (b) the value of C;
   (c) the power factor of the unknown arm.
   Answer: 4 Ω, 0.6 µF, 0.0227

12. A Schering bridge is used to measure the capacitance and parallel loss resistance of a length of coaxial cable. The supply voltage across the bridge is 2.5 kV at 50 Hz. The bridge component values for balance are

   $C_1 = 100$ pF $\quad R_2 = 420$ Ω $\quad C_2 = 1$ µF ± 0.001 µF $\quad R_3 = 160$ Ω ± 0.6 Ω

   Determine:
   (a) the cable capacitance;
   (b) the parallel loss resistance;
   (c) the power loss in the cable.
   Answer: 0.22 nF + 0.3 per cent, 109.2 MΩ, 0.057 W
   (Hint: The parallel loss resistance is given by the expression $(R_p = 1/Rw^2C^2)$.)

13. A Wien bridge similar to the one shown in Fig. 2.26 has to be used in a Wien bridge oscillator to produce a range of test tones between 750 Hz and 4.2 kHz. If $R_2 = R_3 = 10$ kΩ determine the range of C_2 and C_3.
   Answer: 3.79–21.23 nF

14. A full-wave power supply has to be designed with the following parameters:

   Load resistance $= 27$ Ω

   Load voltage $= 12$ V

   Supply voltage $= 240/50$ Hz (r.m.s.)

Calculate the value of filter capacitor suitable for this supply.
Answer: 192 µF

15. A full-wave 450 V_{peak}/50 Hz supply is applied across a resistive load of 9.2 kΩ when the load voltage is 445 V. Calculate the value of capacitor necessary to handle the ripple voltage involved.
Answer: 48.9 µF

16. A 10 V pulse generator produces a pulse having a duration of 5 ms which is applied to the input of an integrator having $R = 1$ kΩ and $C = 0.002$ µF.
(a) Determine the voltage to which the capacitor will charge.
(b) Determine the time it takes for the capacitor to discharge through the source if its internal resistance is 75 Ω.
(c) Sketch the output voltage.
Answer: 10 V, 10.75 µs

17. In a differentiator circuit shown in Fig. 2.47 the rheostat is set to 682 Ω. Determine:
(a) the output waveform under these conditions;
(b) the output waveform when the rheostat is set to 850 Ω;
(c) the effect on the output waveform if the 2.2 µF capacitor is short-circuited.
Answer: (a) The transition is between +2.03 and −12.97 V
        (b) The transition is between +3.02 and −11.98 V

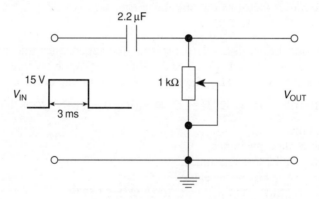

Figure 2.47

18. A timer circuit has to be designed to activate a triggering circuit which in turn will switch on a control system at a particular threshold level of 2.4 V. The situation is shown in Fig. 2.48. The triggering circuit triggers at point A on the output waveform as shown and the circuit has to be designed for delays of 5, 15 and 30 ms. Determine the three settings of VR.
Answer: 42.47 kΩ, 127.42 kΩ, 254.84 kΩ

19. A capacitive displacement sensor has plates which are 30 mm square and 4 mm apart. The ceramic dielectric has to move a distance 25 cm. Determine the changes in capacitance which a bridge will sense (ε_r for ceramic is 1200).
Answer: 67.38–2390 pF

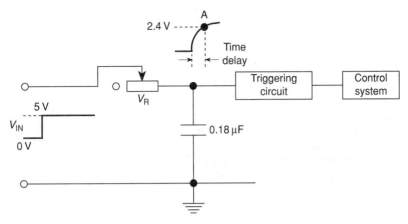

Figure 2.48

20. A linear displacement sensing unit has to be designed using a Wien bridge oscillator and an amplifier as shown in Fig. 2.49. If both the capacitive sensors are
identical and are coupled to two objects causing the same displacement,
determine the frequency available at the output if the moving dielectric is
operational between 5 and 25 mm of travel. The following parameters apply
for the sensors: $A = 900^2$ m, $\varepsilon_r = 5$, $d = 3$ mm, $\varepsilon_0 = 8.854 \times 10^{-12}$ pF.
Answer: 17.133–71.922 kHz

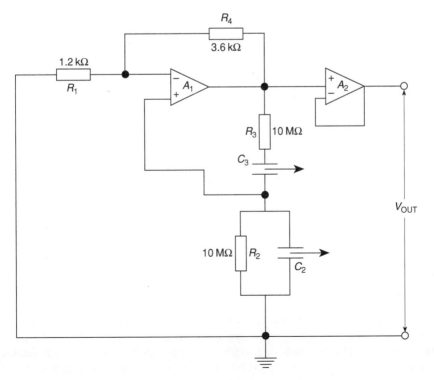

Figure 2.49

3

Discrete and integrated semiconductor physics

3.1 Introduction

The schematic diagram shown in Fig. 3.1 is called a d.c. level converter and is used in special power supplies. This circuit uses many types of semiconductor device, each one fulfilling a particular function. Some of these components are discrete or individual devices, while others use many discrete components etched on to a single chip. Indeed it is possible for the complete circuit of Fig. 3.1 to be fabricated on a single chip, with the exception of the inductor L_1.

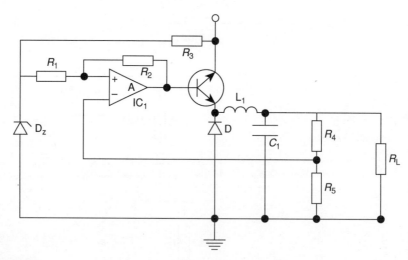

Figure 3.1

In order to understand how these devices function, a knowledge of semiconductor physics is necessary. Only then does the electronic operation of circuits become clear. The function of this chapter is to understand the fundamental principles of semiconductor physics and then apply these to such devices as diodes, transistors and integrated circuits.

3.2 Bohr atom

As was explained in Chapter 1, the classical idea of the atom is of a model consisting of a nucleus with orbiting electrons. However, as the electrons are moving in a circle their velocity is changing and hence they are accelerating. This implies that energy is being radiated, as any object which accelerates emits energy. The orbit of the electron should decay, but this does not appear to be the case. This problem was solved by Niels Bohr in 1913. He postulated the following:

1. Electrons can have only certain orbital radii, which means that they are allowed to have only certain values of energy (energy levels). The allowed energy levels are referred to as stationary states as an electron can remain in an energy level indefinitely without radiating any energy.
2. An electron can absorb electromagnetic radiation in the form of photons, and this causes the electron to jump to an outer orbit (E_2).
3. The electron can spontaneously emit energy in the form of radiation when it falls to a lower orbit (E_1).

These ideas are shown in Fig. 3.2. Note from this diagram that electrons can move randomly and do not need to return to the same orbit they originated from. For example, an electron jumps from E_1 to E_2, but it may fall to E_3. The energy radiated when an electron falls from E_2 to E_1 is given by the equation:

$$E_2 - E_1 = hf \tag{3.1}$$

where f is the frequency and h is known as Planck's constant (6.62×10^{-34} J s). Also

$$f = \frac{v}{\lambda} \tag{3.2}$$

where λ is the wavelength (m) and v is the velocity of electromagnetic energy ($3 \times 10^8 \, \mathrm{m\,s^{-1}}$).

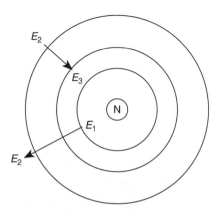

Figure 3.2

The fundamental equation (3.2) is normally used with the photoelectric effect to indicate the movement of particles in metals and semiconductor materials. Two points should be observed here:

1. The more intense the incident radiation the greater the number of moving electrons, i.e. their quantity increases but not their kinetic energy.
2. Before an electron can leave a material it must have a certain energy level, just as molecules acquiring sufficient energy boil off from boiling water. In the case of the photoelectric effect this minimum energy is known as the work function (Φ) and it depends on the material

$$\Phi = hf_0 \tag{3.3}$$

where f_0 is known as the threshold frequency and is the minimum frequency the material can be subjected to before the photoelectric effect occurs. The energy which the electron will have may now be stated as

$$E = hf - \Phi = hf - hf_0 \tag{3.4}$$

When dealing with the photoelectric effect and energy levels the joule is not used as the unit of energy as it is too small. Instead the electron-volt is used and this is equal to the kinetic energy gained by an electron which has been accelerated between two points where the potential difference is 1 V. The work done (E) when a particle having a charge Q moves through a p.d. of V volts is given by

$$E = QV \tag{3.5}$$

The charge on a single electron is 1.6×10^{-19} C and when the p.d. is 1 V then the work done is

$$(1.6 \times 10^{-19}) \times 1 = 1.6 \times 10^{-19} \, \text{J}$$

As the work done is equal to the kinetic energy gained by the electron, then

$$1 \, \text{eV} = 1.6 \times 10^{-19} \, \text{J}$$

Finally, in this section the photon has been mentioned. This is a wave packet and generally the orbiting electron absorbs complete packets or photons. However, when the energy is radiated, all or part of the energy may be given out depending on what part of the electromagnetic spectrum the original photon came from, i.e. the electron may fall to different orbits as shown in Fig. 3.2 and hence radiate the different wavelengths, for example visible, infrared or ultraviolet.

Example 3.1

A light-emitting diode (LED) is connected to a 3 V supply and is capable of emitting light at a wavelength of 620 nm. Determine:

(a) the kinetic energy which the electrons must have;
(b) the work function of the material.

Solution

(a) $E_K = (1.6 \times 10^{-19}) \times 3$

(b) $\Phi = \dfrac{h\upsilon}{\lambda} - E = \dfrac{\left(\dfrac{6.62 \times 10^{-34} \times 3 \times 10^8}{620 \times 10^{-9}}\right) - 1.6 \times 10^{-19}}{1.6 \times 10^{-19}} = -1$

The negative answer means that electrons do not have enough energy to escape, hence the LED is functioning outside its spectral response.

Example 3.2

It is required to calculate the wavelength of radiation emitted as a result of an electron transition from 0.85 to −3.39 eV.

Solution

$$E_2 = 0.85\,\text{eV} = -0.85 \times 1.6 \times 10^{-19} = -1.36 \times 10^{-19}$$

$$E_1 = -3.39\,\text{eV} = -3.39 \times 1.6 \times 10^{-19} = -5.424 \times 10^{-19}$$

From equation (3.2)

$$E_2 - E_1 = \frac{h\upsilon}{\lambda}$$

$$\therefore \lambda = \frac{h\upsilon}{(E_2 - E_1)} = \frac{6.62 \times 10^{-34} \times 3 \times 10^8}{(-1.36 \times 10^{-19}) - (-5.424 \times 10^{-19})}$$

$$= \frac{19.86 \times 10^{-26}}{4.064 \times 10^{-19}} = 488.7\,\text{nm}$$

3.3 Basic semiconductor formation

Modern chip technology is cheap and reliable and it is now possible to etch most components on to a silicon substrate. A resistor is a conductor whose length and cross-sectional area control its value while a capacitor may have its value changed by varying the area of the plates and the thickness of the insulator between the plates. Diodes and transistors are also fabricated by more complex methods and this enables them to be used in a multitude of complex circuits to fulfil different functions. A useful function of the diode is to etch it in a particular way so that it acts as an open circuit to isolate different parts of a circuit.

It is not the purpose of this chapter to look at the complex fabrication processes involved in the semiconductor industry, but an understanding of the basic chemistry and physics is necessary before graduating to energy levels and circuit applications.

After oxygen, silicon is the next most prolific element. Silicon is a solid, and one of the important properties of a solid is its ability to conduct electrons. The resistance of a number of different solids can be shown on a scale as shown in

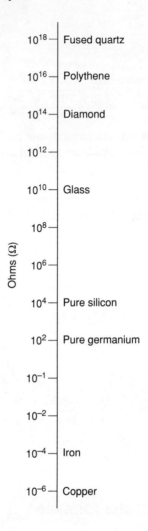

Figure 3.3

Fig. 3.3. This figure shows the huge diversity in material resistance. Materials at the top are used as insulators while those at the bottom are used as conductors. Silicon appears in the middle and hence is called a semiconductor. It is because of its position that it is sensitive to the presence of certain impurities called dopants in the industry. The addition of extremely small quantities of dopant can improve conductivity by factors of 1000.

The elementary two-dimensional picture of pure silicon is shown in Fig. 3.4(a). Each silicon atom has four electrons in its outer shell with spaces for another four. When the atom is surrounded by similar atoms as in Fig. 3.4(b), its four electrons are shared with neighbouring atoms which in turn share their electrons. Consequently each nucleus is effectively surrounded by eight electrons. This structure is very stable and as a result it is difficult for electron flow to take place.

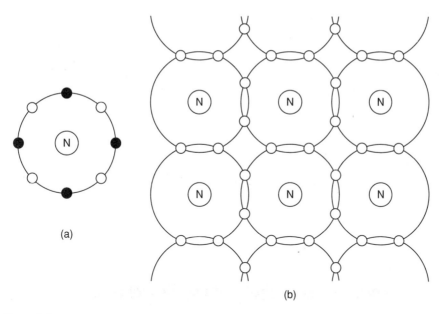

(a)

(b)

Figure 3.4

In order to cause current flow it is necessary to modify this stable structure and this is done by doping the silicon with dopants such as boron and phosphorus. Figure 3.5 shows the outer electron shell of these two elements compared to silicon. If each of these impurities is added to separate silicon slices the situation is as shown in Fig. 3.6. In Fig. 3.6(a) a boron impurity has been added so that a 'hole' occurs in the lattice. This hole means that there is an excess positive charge on the adjacent silicon nucleus and so the hole can move through the silicon lattice. For this reason boron is called a p-type dopant. Figure 3.6(b) shows the effect of doping the silicon with phosphorus. In this case there is one remaining electron and it is available to take part in current flow. Phosphorus is therefore called an n-type dopant. This is the birth of semiconductor fabrication and when a silicon crystal is half doped with p-type dopants and the other half doped with n-type dopants a basic semiconductor building block is constructed upon which all others are built.

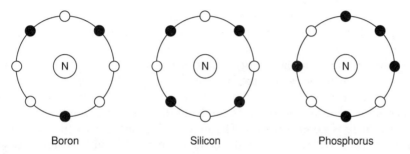

Boron   Silicon   Phosphorus

Figure 3.5

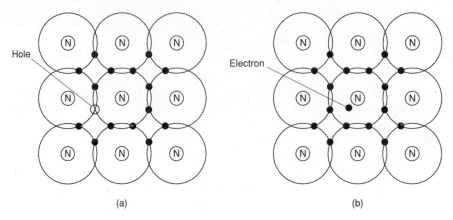

(a)   (b)

Figure 3.6

3.4 Energy levels as applied to semiconductors

The electrons in an isolated atom have a well-defined set of energy levels and all atoms of a given element have a specific set of energy levels. When two identical atoms are close together their electrons are influenced by the combined electric fields of the two atoms and a previously simple energy level splits into two levels. This mainly applies to the outer electron shell and, when large numbers of atoms exist as in Fig. 3.6, the energy levels continue to split and form bands. Each band contains a large number of levels and these are so close together that a continuous range of energies is available. The energy bands are separated by gaps in which there are no energy levels.

If an electron has to flow through a crystal lattice as in Fig. 3.6(b) it must be accelerated by a p.d. so that it has sufficient energy to be raised to a higher energy level (orbit). Consequently, a material can conduct electrons only if some of its electrons are in a band which does not have a full complement of electrons. If this is not the case then higher energy levels have to be considered which may be prohibitive.

For a moment let us consider single energy levels or states as shown in Fig. 3.7. This diagram illustrates the points already mentioned. The electron can exist at several levels, the gap between any two energy levels is called the energy gap and this has to be overcome by the electron and finally if an electron reaches the zero energy level it will leave the influence of the atom completely and the atom becomes ionized.

Figure 3.8 shows actual energy bands due to level splitting. Figure 3.8(a) illustrates an insulator which has a large energy gap, hence electrons do not have enough energy to bridge the gap. If they do then the insulator has broken down. Figure 3.8(b) shows the band diagram for a conductor and in this case there is no energy gap. The two bands overlap with the result that many electrons are available in the conduction band. Figure 3.8(c) shows a pure piece of silicon (intrinsic) at 0 K. The valence band is completely full

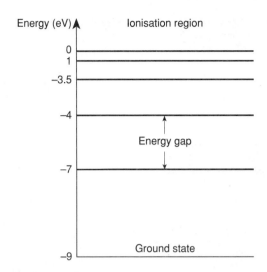

Energy (eV) — Ionisation region
0
1
−3.5
−4
Energy gap
−7
Ground state
−9

Figure 3.7

while the conduction band is totally empty. This is similar to the insulator except that the energy gap is a lot narrower. As the temperature is increased some electrons acquire enough energy to jump the gap. This is the source of leakage current in a semiconductor and it should be kept low if the semiconductor is to function properly. Figure 3.8(d) shows this.

As was mentioned in section 3.3, dopants are generally added to intrinsic silicon turning it into an extrinsic semiconductor. Figure 3.8(e) shows the band diagram for an n-type material. The donor levels are just below the conduction band and a lot less energy (0.1 eV) is now required to raise these levels to the conduction band. Similarly, the addition of an acceptor impurity to an intrinsic semiconductor creates extra energy levels just above the valence band as in Fig. 3.8(f). The acceptor levels are occupied by thermally excited electrons from the valence band and this leaves a large number of holes in the valence band for increased conductivity.

3.5 Significance of the Fermi level and the Boltzmann principle

Electrons acquire different energies in different atoms and some of these electrons may move to higher energy states than their neighbour. There are many energy states available for these electrons, but the electron which reaches the highest energy state at 0 K establishes a level called the Fermi level. This level is important when carriers diffuse across a barrier such as that created when a p- and n-type material form a junction.

A well-known principle called the Boltzmann principle deals with the equilibrium of two energy states when they are combined. This is shown in Fig. 3.9. This situation occurs when a steady state is reached. Under equilibrium

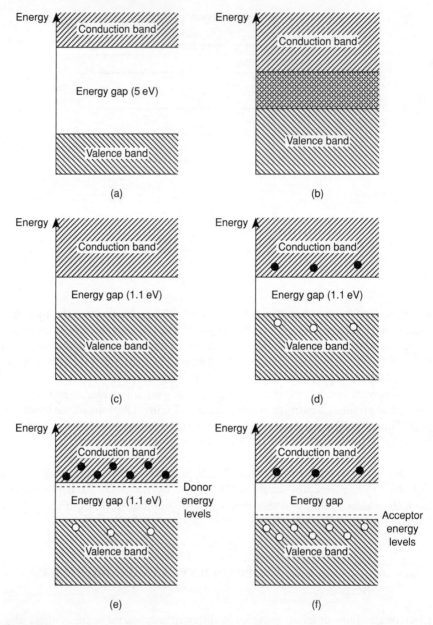

Figure 3.8

there is a steady flow of energy from E_1 to E_2. The same principle can be applied to n- and p-type materials when they are fused together. In Fig. 3.10 equilibrium is achieved when the two Fermi levels in the n- and p-type materials align. Under this condition electrons will flow in the direction shown as there is virtually a difference in potential energy. This structure forms the basis of the p–n junction diode considered in the next section.

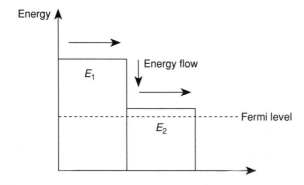

Figure 3.9

3.6 p–n junction diode

When equilibrium occurs in a p–n junction a depletion layer or junction barrier is formed as shown in Fig. 3.10(b). The junction barrier is an ionized region which prevents further conduction due to an opposing electric field. Typically 0.7 V is taken as the p.d. across the junction.

When an external voltage is applied across the p–n junction diode it may be connected in two ways as shown in Fig. 3.10(c). In Fig. 3.10(a) the diode is forward biased and large electron and hole flows occur, i.e. the majority carriers are high in both directions. In Fig. 3.10(b) the diode is reverse biased

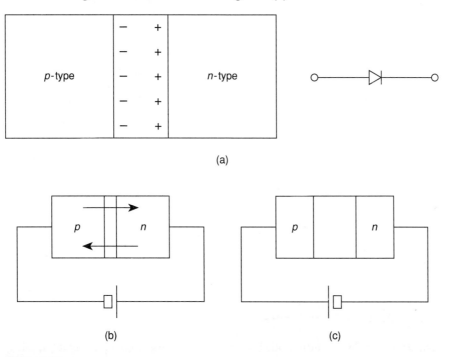

Figure 3.10

and virtually no flow occurs except for thermal excitation (leakage current). The depletion region is narrow for forward biasing and wide for reverse biasing.

3.7 Zener diode

There are many types of diode used in electronic circuits, but for the moment one important one will be considered, namely the zener diode. The characteristics shown in Fig. 3.11 will clarify the function of this diode. The p–n junction diode generally operates in the forward-biased region and conducts when the voltage across it exceeds 0.7 V approximately. This type of diode is not designed for high reverse voltage which would destroy it. However, the zener diode is designed to operate in the reverse bias region. Figure 3.11 shows the effect of applying a large reverse voltage to a junction diode. There is a large increase in current as the reverse voltage is increased beyond the breakdown voltage. This is called the zener voltage and the breakdown is a result of the electric field across the depletion layer becoming high enough to pull electrons away from their nuclei and so produce a large number of electron–hole pairs. This causes a massive increase in the reverse current. Zener diodes have high doping levels in order to produce these large reverse currents which are due to minority carriers.

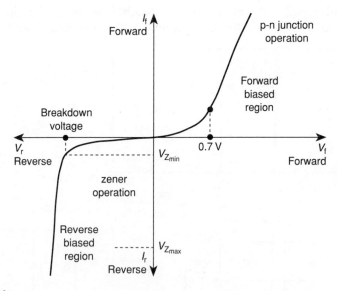

Figure 3.11

3.8 Diode applications

The general p–n junction diode may be used for large and small signal applications and because of this they have different encapsulations and ratings. Typical

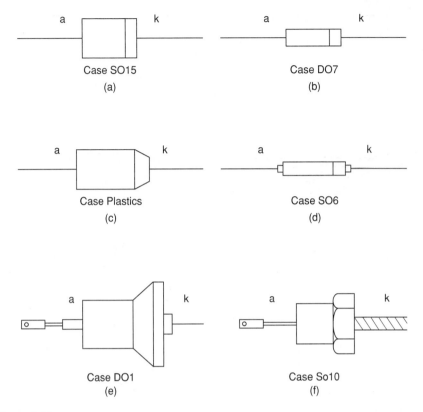

a   Case SO15
(a)

a   Case DO7
(b)

a   Case Plastics
(c)

a   Case SO6
(d)

a   Case DO1
(e)

a   Case So10
(f)

Figure 3.12

packages are shown in Fig. 3.12. As with all active components, diode data sheets are provided by the manufacturers. Data sheets can be confusing due to the plethora of information provided and the parameter symbols used.

Example 3.3

Determine the current for each of the circuits shown in Fig. 3.13, assuming that the forward-biased resistance is $25\,\Omega$ and the reverse-bias resistance is $100\,\text{M}\Omega$.

Solution

(a)
$$I_f = \frac{10}{75 + 25} = 100\,\text{mA}$$

(b)
$$I_r = \frac{500}{100 + 1} = 4.95\,\mu\text{A}$$

(c)
$$I_r = \frac{12}{100 + 1} = 0.12\,\mu\text{A}$$

(d)
$$I_r = \frac{5}{50 + 25} = 66.7\,\mu\text{A}$$

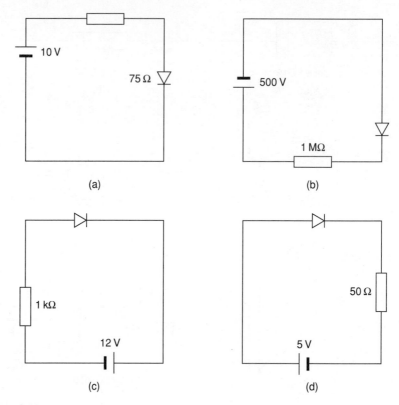

Figure 3.13

3.9 Power supplies

A power supply is necessary to drive instrumentation and electronic systems. The mains supply (240 V_{rms}, 50 Hz) is converted from a.c. to some d.c. value for ongoing processing. After d.c. conversion filtering is required, followed by some form of regulation and stabilization. In addition some form of automatic overload and overvoltage protection is required. A simple block diagram of the components required for a standard power supply is shown in Fig. 3.14.

Figure 3.14

Half-wave rectifier

The simplest type of power unit is shown in Fig. 3.15 which consists of a simple diode and transformer. The transformer steps the voltage up or down depending on the desired output voltage. The diode only conducts during the

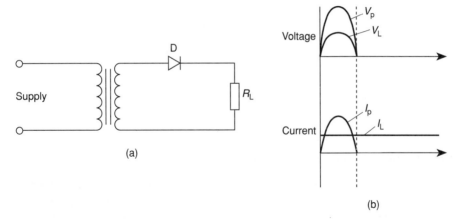

Figure 3.15

positive half-cycle of the input voltage when it is forward biased. During the negative half-cycle the diode is reverse biased and is cut off. The output voltage therefore consists of a series of unidirectional pulses at the supply frequency of 50 Hz.

Several points should be noticed about this supply:

1. The diode should have a voltage rating of twice the peak value of the voltage across the transformer. This is known as the peak inverse voltage (PIV). Hence this may be given as

$$PIV = 2V_p \qquad (3.6)$$

2. The average d.c. or load voltage is related to the r.m.s. voltage by the following relationship:

$$V_L = 0.45\,V_{rms} \qquad (3.7)$$

3. The load, average or d.c. current is related to the peak current by the following relationship:

$$I_L = 0.318 I_p \qquad (3.8)$$

4. The power rating of the transformer is given in terms of the r.m.s. values of the secondary voltage and current

$$P_{rms} = V_{rms} I_{rms} \qquad (3.9)$$

5. Remember:

$$I_p = 1.414 I_{rms} \qquad (3.10)$$
$$V_p = 1.414\,V_{rms} \qquad (3.11)$$

Full-wave rectifier

Figure 3.16 shows the full-wave rectifier circuit. It consists of two diodes each of which conducts over alternate half-cycles of the input voltage of the primary.

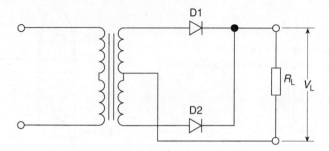

Figure 3.16

This is achieved by means of a centre-tapped transformer which provides 180° phase shift between the top and bottom of the secondary. So during the positive half-cycle the top end of the transformer is positive and the bottom end negative. During the negative half-cycle the roles are reversed. The output voltage thus consists of a series of unidirectional pulses, but this time the frequency of these pulses is at twice the mains frequency, i.e. 100 Hz.

As before, the diodes have to be capable of withstanding twice the peak value of the secondary voltage when they are reverse biased. The following relationships once again hold:

1.
$$\text{PIV} = 2V_\text{p} \tag{3.12}$$

2.
$$V_\text{L} = 0.9\,V_\text{rms} \tag{3.13}$$

3.
$$I_\text{L} = 0.636 I_\text{p} \tag{3.14}$$

Full-wave bridge rectifier

The disadvantages of the two previous supplies is the PIV rating of the diodes and the use of a centre-tapped transformer which can be expensive. Both these problems can be eliminated by using a bridge configuration as shown in Fig. 3.17. In this circuit pairs of diodes conduct during each half-cycle.

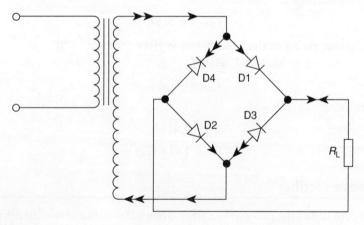

Figure 3.17

During the positive half-cycle D_1 and D_2 conduct the path being shown by the single arrow. During the negative half-cycle D_3 and D_4 conduct, and this path is shown by the double arrow. As before, the load will receive pulsating d.c. voltage at 100 Hz. This is by far the most frequently used configuration and the four diodes are normally encapsulated in a four-terminal block for easy printed circuit board mounting. Note a disadvantage of this circuit is that it is not possible to simultaneously earth one side of the power supply and one side of the output, otherwise part of the bridge would be short-circuited.

Example 3.4

A half-wave rectifier supply circuit supplies 150 mA to a 500 Ω load. Calculate:

(a) the d.c. load voltage;
(b) the r.m.s. voltage of the transformer;
(c) the minimum peak inverse rating of the diode.

Solution

(a)
$$V_L = R_L I_L = \frac{150 \times 5 \times 10^2}{10^3} = 75 \text{ V}$$

(b)
$$V_L = \frac{V_p}{\pi}$$
$$\therefore V_p = \pi V_L = 3.14 \times 75 = 235.5 \text{ V}$$

(c)
$$\text{PIV} = 2 \times 235.5 = 471 \text{ V}$$

Example 3.5

A half-wave supply with a resistive load of 2.2 kW is shown in Fig. 3.18. Determine the PIV rating and the average forward current of the diode.

Solution
The secondary voltage is 240 V_{rms} as the turns ratio is 1:1. Hence the load or d.c. voltage is

$$V_L = \frac{V_p}{\pi} = \frac{240 \times 1.414}{\pi} = 108 \text{ V}$$

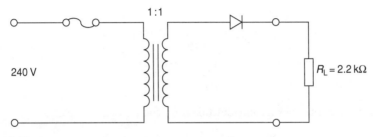

Figure 3.18

The PIV is

$$PIV = 2 \times V_p = 2 \times 240 \times 1.414 = 679 \text{ V}$$

The average forward current (I_F) is the current through the load

$$\therefore I_F = \frac{108}{2.2 \times 10^3} = 49 \text{ mA}$$

Note this ignores the 0.7 V drop across the diode. At lower d.c. voltages this would have to be considered.

Example 3.6

A full-wave centre-tapped supply having a capacitive input filter has to provide a load of 1 kΩ at 350 V d.c. If $C = 120 \, \mu F$, determine:

(a) the r.m.s. voltage of the transformer required;
(b) the average current taken by each rectifier;
(c) the PIV rating of each diode.

Solution

(a) From (2.52)

$$V_L = \left(1 - \frac{T}{R_L C}\right) V_{p,\text{in}}$$

$$\therefore V_{p,\text{in}} = \frac{V_L}{1 - (T/R_L C)}$$

$$= \frac{350/(1 - 0.01)}{10^3 \times 120 \times 10^{-6}} = 381 \text{ V}$$

Also

$$V_{p,\text{in}} = \frac{V_{p,\text{sec}}}{2} - 0.7$$

$$\therefore \frac{V_{p,\text{sec}}}{2} = V_{p,\text{in}} + 0.7 = 381 + 0.7$$

$$\therefore V_{p,\text{sec}} = 763.4$$

$$\therefore V_{\text{rms}} = \frac{763.4}{1.414} = 539.9 \text{ V}$$

(b)
$$I_{dc} = \frac{V_{dc}}{R_L} = \frac{350}{1000} = 350 \text{ mA}$$

Therefore, the average current taken by each diode is 175 mA.

(c)
$$PIV = 2V_{p,\text{in}} = 2 \times 381 = 762 \text{ V}$$

Example 3.7

A full-wave bridge rectifier supply with a capacitive input filter has to supply 6 V d.c. load having a resistance of $1.2 \, \text{k}\Omega$. The diodes have to withstand a surge current (I_{FSM}) of 30 A. Determine:

(a) the PIV of the diodes
(b) the value of the surge resistor
(c) the value of C.

Solution

(a)
$$\text{PIV} = \text{the peak output voltage } (V_{po})$$

$$V_{po} = V_{p,sec} - 1.4$$

$$V_{p,sec} = \frac{V_L \times \pi}{2} = \frac{6 \times 3.14}{2} = 9.42 \, \text{V}$$

$$\therefore \ V_{po} = 9.42 - 1.4 = 8.02 \, \text{V}$$

(b)
$$V_L = \left(1 - \frac{T}{R_L C}\right) V_{po}$$

Transposing gives

$$C = \frac{T}{R_L(1 - V_L/V_{po})}$$

$$= \frac{0.01}{1.2 \times 10^3 (1 - 6/8.02)} = 33.1 \, \mu\text{F}$$

(c)
$$R_s I_{FSM} = V_{p,sec} - 1.4$$

$$\therefore \ R_s = \frac{V_{p,sec} - 1.4}{I_{FSM}} = \frac{9.42 - 1.4}{30} = 0.27 \, \Omega$$

3.10 Diode limiting circuits

Diode circuits are sometimes used to clip off portions of a signal voltage above or below certain levels. These are known as limiters or clippers. Clipping can be accomplished by overdriving an amplifier, but biased diodes are a more efficient way of accomplishing this. Figures 3.19(a) and (b) show two simple parallel clipper configurations with the associated output waveforms. Figure 3.19(a) limits the positive part of the input a.c. voltage. As the input signal becomes positive the diode is forward biased, and since the cathode is at ground potential the anode cannot exceed 0.7 V. When this input voltage falls below 0.7 V the diode reverse biases and appears as an open circuit and the negative part of the waveform appears at the output, but with a magnitude given by

$$V_o = \left(\frac{R_L}{R_L + R_s}\right) V_i \qquad\qquad (3.15)$$

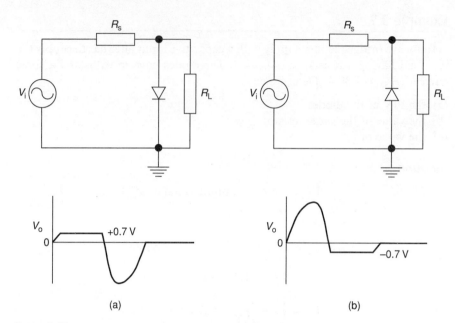

(a)   (b)

Figure 3.19

If the diode is now reversed as in Fig. 3.19(b) the negative part of the input is clipped. When the diode is forward biased during the negative part of the input the cathode is held at $-0.7\,\text{V}$, and when the input goes above $-0.7\,\text{V}$ the diode is no longer forward biased and the voltage during the positive half-cycle is produced.

For certain applications such as the input to a voltage amplifier it may be necessary to alter the clipping level. This is achieved by circuits similar to those found in Figs 3.20(a) and (b). Notice the output waveform in each case. The bias voltage V_B may be derived from a voltage divider circuit or reference resistor.

Example 3.8

A clipper circuit has to be designed so that the positive cycle of a 1 V peak-to-peak input a.c. voltage is clipped at 2.7 V and the negative cycle swings to 4 V. Design this circuit if the load is $600\,\Omega$.

Solution
The circuit required is shown in Fig. 3.21. The voltage at the anode of the diode after the diode conducts is limited to

$$V_B + 0.7 = 2.7$$

$$\therefore\ V_B = 2\,\text{V}$$

If the negative cycle is limited to 4 V, then

$$V_o = \left(\frac{R_L}{R_L + R_s}\right) V_i$$

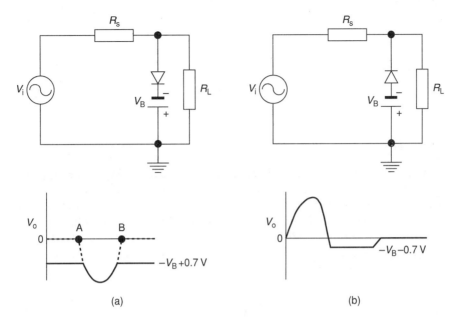

(a)   (b)

Figure 3.20

Arranging gives

$$R_s = \frac{V_i R_L - V_o R_L}{V_o}$$

$$= \frac{6 \times 600 - 4 \times 600}{4}$$

$$\therefore\ R_s = 300\,\Omega$$

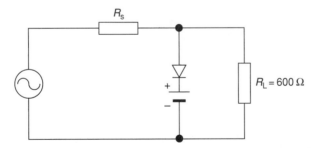

Figure 3.21

3.11 Diode clamping circuits

Clamping circuits are used to shift an a.c. waveform up or down by adding a d.c. level which is equal to the positive or negative value of the a.c. signal. These circuits are also called level shifters or d.c. restorers.

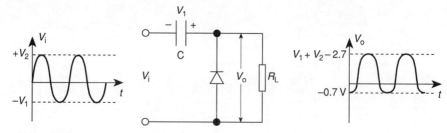

Figure 3.22

Figure 3.22 shows a typical clamping circuit. When the input goes negative the diode is forward biased and hence the capacitor charges rapidly to the peak voltage V_1. The charge time constant is very small because the forward resistance of the diode is small, hence the capacitor has the voltage shown across it. Generally the CR_L time constant is large and is taken as 10 times the period T of the input frequency. If Kirchhoff's voltage law is used round the capacitor–diode loop then the voltage across R_L is

$$V_L = V_1 + V_2 - 0.7 \qquad (3.16)$$

The voltage is thus shifted up by an amount equal to the negative peak voltage, whereas if the diode was reversed the voltage would be shifted down by an amount equal to the positive peak voltage.

Example 3.9

Determine the output voltages in each case in Fig. 3.23.

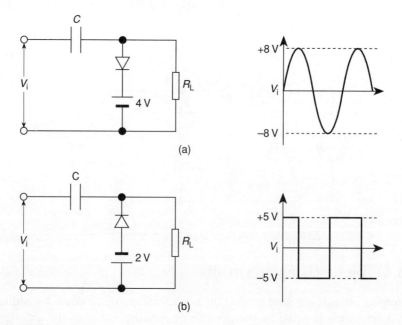

Figure 3.23

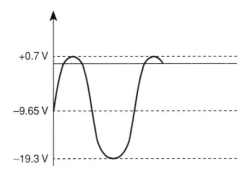

Figure 3.24

Solution

(a) Using Kirchhoff's voltage law gives (Fig. 3.24)

$$V_L = V_1 + V_2 + 4 - 0.7$$
$$= 8 + 8 + 4 - 0.7 = 19.3 \text{ V}$$

(b) Using Kirchhoff's voltage law gives (Fig. 3.25)

$$V_L = V_1 + V_2 + 2 - 0.7$$
$$= 5 + 5 + 2 - 0.7 = 11.3 \text{ V}$$

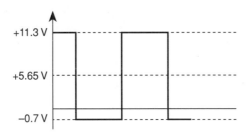

Figure 3.25

3.12 The zener as a regulator

In section 3.8 the zener diode was discussed. Recall that this diode has the capability of being operated in the reverse bias mode, but maintains a constant reference voltage. The power supply circuits considered in section 3.10 are referred to as unregulated supplies as they do not take into account varying input voltages such as ripple voltage and variations due to a changing load. Both these conditions may be overcome in simpler supplies by using a zener stabilizer. Larger current supplies generally use three or five terminal regulators. Note that the characteristic shown in Fig. 3.11 may be operated between $V_{z(min)}$ and $V_{z(max)}$. If the diode operates below $V_{z(min)}$ then stabilization is

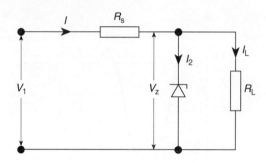

Figure 3.26

lost as the voltage now varies with variation of current. Similarly there is a limit to $V_{z(max)}$ because of the power requirements of the zener.

Figure 3.26 shows a simple voltage stabilizer. This circuit consists of a zener diode with its cathode connected to the positive pole of the unstabilized power supply via resistor R. The supply voltage V_1 is greater than the zener diode break-down voltage V_z, hence it follows that the voltage across R is $(V_1 - V_2)$ volts. The output voltage V_2 across the load R_L is equal to V_z. If the supply voltage V_1 alters in value then the p.d. across R_L alters to compensate for the change and the output voltage remains constant. If the current in the resistor R is I, then

$$R = \frac{V_1 - V_z}{I} \tag{3.17}$$

When the load resistor R_L is disconnected the diode must be capable of carrying the current I given by equation (3.17). So the power rating of the diode sets an upper limit for the value of I:

$$I = \frac{P_z}{V_z} \tag{3.18}$$

where P_z is the power rating of the diode.

Circuit regulation

There are two considerations to be taken into account when designing a power supply using a zener stabilizer:

- line regulation
- load regulation.

Line regulation is generally concerned with varying input voltage from the unregulated section of the supply due to ripple or other factors, while load regulation refers to varying load conditions. Note at this stage that I varies with variations in V_1, but I_L only varies with change in R_L.

Line regulation

This type of discussion is best illustrated by means of an example.

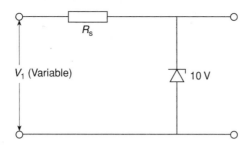

Figure 3.27

Example 3.10

Consider the circuit shown in Fig. 3.27 in which a zener is used which has the following characteristics:

$$I_{z(min)} = 4\,mA \qquad I_{z(max)} = 40\,mA$$

Find the minimum and maximum values of input voltage V_1 which can be used with this zener.

Solution

If the current in the zener is at its minimum value, then

$$V_r = 4 \times 10^{-3} \times 10^3 = 4\,V$$

$$V_{1(min)} = V_r + V_z = 4 + 10 = 14\,V$$

If the current in the zener is at its maximum value, then

$$V_r = 40 \times 10^{-3} \times 10^3 = 40\,V$$

$$V_{1(max)} = V_r + V_z = 40 + 10 = 50\,V$$

Example 3.11

Determine the minimum and maximum input voltages that can be regulated by the diode in Fig. 3.28 assuming the diode has the following characteristics:

$$I_{z(min)} = 1\,mA \qquad I_{z(max)} = 15\,mA \qquad V_z = 5.1\,V \qquad R_z = 10\,\Omega$$

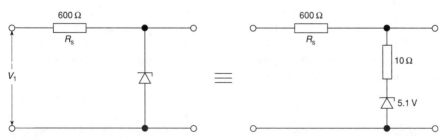

Figure 3.28

For $I_{z(min)}$

$$V_o = V_z + I_{z(min)} R_z$$

$$= 5.1 + 1 \times 10^{-3} \times 10 = 5.11 \text{ V}$$

$$V_{1(min)} = I_{z(min)} R + V_o$$

$$= 1 \times 10^{-3} \times 600 + 5.11 = 5.17 \text{ V}$$

For $I_{z(max)}$

$$V_o = V_z + I_{z(max)} R_z$$

$$= 5.1 + 15 \times 10^{-3} \times 10 = 5.25 \text{ V}$$

$$V_{1(max)} = I_{z(max)} R + V_o$$

$$= 15 \times 10^{-3} \times 600 + 5.25 = 14.25 \text{ V}$$

Load regulation

Consider the situation where a load is now connected to the zener stabilizer as in Fig. 3.29. When the output terminals are open $R_L = $ infinity, the load current is zero and all of the current is through the zener. As R_L is decreased, I_L goes up and I_z goes down. The zener diode continues to regulate until I_z reaches its minimum value $I_{z(min)}$. At this point the load current is maximum and the load resistance is minimum.

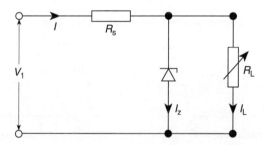

Figure 3.29

Example 3.12

Determine the minimum and maximum load currents for which a zener diode will maintain regulation if it is supplied with a regulated supply of 24 V and the series resistor is 500 Ω. Also, determine the minimum value of R_L. The zener has the following characteristics:

$$V_z = 12 \text{ V} \qquad I_{z(min)} = 3 \text{ mA} \qquad I_{z(max)} = 90 \text{ mA} \qquad R_z = 0$$

Solution

Suppose that the load is open circuited, i.e. $I_L = 0$, then I_z is maximum and equal to the total current I:

$$I_z = \frac{V_1 - V_z}{R_s} = \frac{24 - 12}{500} = 24\,\text{mA}$$

Since this is much less than $I_{z(\text{max})}$, 0 A is an acceptable minimum value for I_L. The maximum value of I_L occurs when I_z is minimum:

$$I_{L(\text{max})} = I - I_{z(\text{min})} = 24 - 3 = 21\,\text{mA}$$

The minimum value of R_L is

$$R_{L(\text{min})} \frac{V_z}{I_{L(\text{max})}} = \frac{12}{21 \times 10^3} = 571\,\Omega$$

Example 3.13

Design a voltage reference circuit which uses a 5 V, 400 mW zener diode. If the supply voltage is $11\,\text{V} \pm 2\,\text{V}$ calculate the value of the resistor R_S. Find the lowest value of load resistance that can be connected before the regulator fails to function correctly.

Solution
The maximum allowable current in the diode is

$$\frac{P_z}{V_z} = \frac{400 \times 10^{-3}}{5} = 80\,\text{mA}$$

The resistor R_S must limit I to this value when the maximum value of V_1 is applied and $I_L = 0$:

$$R_S = \frac{V_{1(\text{max})} - V_z}{I_{(\text{max})}} = \frac{13 - 5}{80 \times 10^{-3}} = 100\,\Omega$$

A minimum value for the load resistor R_L occurs when $I_z = 0$ and the circuit fails to act as a regulator. At this point $I_L = I$ and V_1 has its minimum value $V_{1(\text{min})}$.

3.13 Bipolar junction transistor

The bipolar junction transistor (BJT) is so called because it depends on holes and electrons for its operation. This device is effectively two diodes sandwiched together with appropriate biasing.

From Fig. 3.30 it will be seen that there are two types of BJT, i.e. n–p–n and p–n–p. Both are considered at this point but the n–p–n is more generally used and will be analysed for much of the discussions which follow. Figure 3.30 shows the physical construction as well as the symbol used for each type of transistor. The base region is lightly doped and very thin compared to the heavily doped emitter and the moderately doped collector regions. The proper bias arrangements are also shown with the base–emitter junction being forward

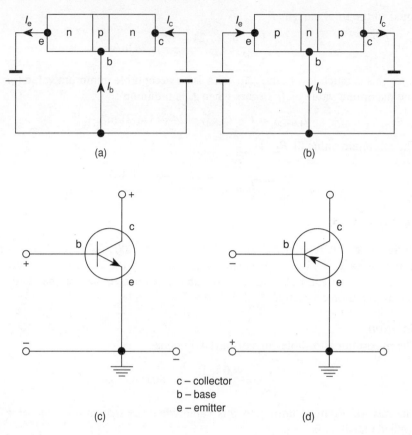

Figure 3.30

biased and the base–collector region being reverse biased. The forward bias from base to emitter narrows the base–emitter depletion region and the reverse bias applied to the base–collector widens the base–collector region. The heavily doped n-type emitter region has an abundance of conduction band electrons that diffuse easily through the forward bias base–emitter junction into the p-type base region. The base region is lightly doped and very thin so that it has a limited number of holes. Thus only a small amount of all the electrons flowing through the base–emitter junction can combine with the available holes in the base and the few recombined electrons flow out of the base lead as valence electrons, forming the small base electron current.

Most of the electrons flowing from the emitter into the thin, lightly doped base region do not recombine, but diffuse into the base–collector depletion region where they are attracted through the reverse-biased base–collector junction by the electric field caused by the positive and negative ions in the junction. The electrons then move through the collector region, and hence towards the positive terminal of the supply. Note that the arrow on the emitter of the transistor symbol indicates hole flow and establishes the type of BJT.

The transistor can be used in many circuits in its discrete form or else, as is increasingly done, in an integrated form on a chip. Irrespective of its configuration there are two fundamental functions of a transistor which will be examined in this chapter, namely

- the transistor as an electronic switch;
- the transistor as a small signal voltage amplifier.

3.14 Transistor as a switch

Figure 3.31 illustrates the basic operation of the transistor as a switching device. Notice the following points concerning the above circuits:

- R_B is a current-limiting resistor included to protect the base of the transistor.
- R_C is called the collector load and is necessary to develop a voltage caused by the collector current.
- As the transistor is n–p–n the supply voltage (V_{CC}) is connected as shown, i.e. the collector positive the emitter negative.

Figure 3.31(a) shows the transistor cut-off because the base–emitter junction is not forward biased and so no current is drawn by the transistor (no collector current). This is effectively an open circuit as indicated by the equivalent switch circuits.

Figure 3.31(b) shows a voltage (V_{BB}) applied between the base–emitter region so that the base–emitter and base–collector junctions are now forward biased. The base current is made large enough to cause the collector current to reach saturation. In this condition there is ideally a short circuit between collector and emitter as indicated by the equivalent switch circuit.

The conditions for cut-off and saturation are given below.

$$\text{Cut-off } V_{CE} = V_{CC} \quad \text{(where } V_{CE} = \text{collector–emitter voltage)} \qquad (3.19)$$

$$\text{Saturation } I_{C(SAT)} = \frac{V_{CC}}{R_C} \qquad (3.20)$$

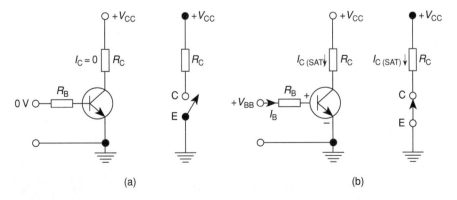

(a)   (b)

Figure 3.31

Hence at cut-off the supply voltage ideally appears across the transistor while at saturation most of the voltage appears across R_C.

3.15 Beta factor (β_{dc}) or d.c. current gain

An important parameter in transistor operation is the d.c. current gain of the transistor and it is given as

$$\beta_{dc} = \frac{I_C}{I_B} \tag{3.21}$$

It is generally designated as h_{FE} on transistor data sheets and typical values are 20–600.

Example 3.14

A sensor circuit connected to a pipeline is designed to emit a symmetrical square wave output with a period of 4 ms. This is connected remotely to a monitor circuit similar to Fig. 3.32. If the light-emitting diode (LED) requires 10 mA to switch it on, determine the amplitude of the square wave input necessary to saturate the transistor given that $\beta_{dc} = 80$ and the LED drops 0.6 V on forward bias.

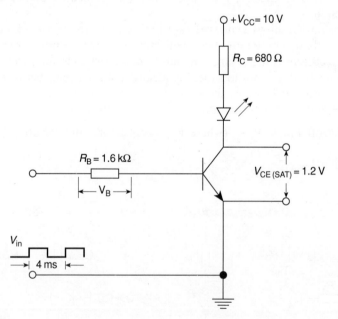

Figure 3.32

Solution

$$I_{C(SAT)} = \frac{V_{CC} - V_{CE(SAT)}}{R_C} = \frac{10 - 1.2}{680} = 12.94\,\text{mA}$$

The minimum value of base current will occur when the transistor is saturated:

$$\therefore \; I_{B(min)} = \frac{I_{C(SAT)}}{\beta_{dc}} = \frac{12.94}{80} = 0.16 \, \text{mA}$$

$$I_B = \frac{V_B}{R_B} = \frac{V_{in} - V_{BE}}{R_B} = \frac{V_{in} - 0.7}{1.6 \times 10^3}$$

$$\frac{0.16 \times 1.6 \times 10^3}{10^3} = V_{in} - 0.7$$

$$\therefore \; V_{in} = \frac{0.16 \times 1.6 \times 10^3}{10^3} + 0.7 = 0.956 \, \text{V}$$

The monitor requires at least 3.26 V to cause it to flash every 2 s.

Example 3.15

The relay in Fig. 3.33 is used to switch a control circuit in such a way that it is activated for 20 s and inhibited for 5 s. If the input voltage has to be at least 5.2 V to switch the transistor on, determine the value of R_B which will enable the circuit to operate. The following conditions apply:

$$\beta_{dc} = 500 \qquad V_{CE} = 0.2 \, \text{V} \qquad r = \text{coil d.c. resistance} = 300 \, \Omega$$

$$I_B = 2I_{B(min)} \quad \text{(to ensure switching)}$$

Solution

$$I_{C(SAT)} = \frac{V_{CC} - V_{CE(SAT)}}{r} = \frac{11.8}{300} = 39.3 \, \text{mA}$$

$$\therefore \; I_{B(min)} = \frac{I_{C(SAT)}}{\beta_{dc}} = \frac{39.3}{500 \times 10^3} = 78.6 \, \mu\text{A}$$

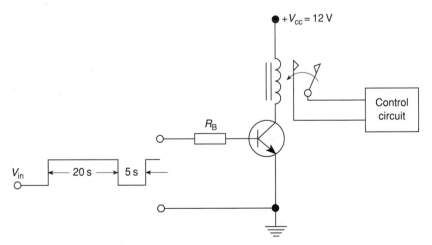

Figure 3.33

$$\therefore\ I_B = 157.2\,\mu\text{A}$$

$$I_B = \frac{V_{\text{in}} - 0.7\,\text{V}}{R_B}$$

$$\therefore\ R_B = \frac{V_{\text{in}} - 0.7}{I_B} = \frac{4.5 \times 10^6}{157.2} = 28.63\,\text{k}\Omega$$

3.16 Transistor as an amplifier

Figure 3.30 showed how d.c. bias was achieved by using two batteries. This is not done in practice. Instead, a widely used biasing method is the voltage divider method which gives good stability using a single supply. This is shown in Fig. 3.34. Note that if the base current is much smaller than the current through R_2 the bias circuit can be considered a voltage divider circuit consisting of R_1 and R_2. If, however, I_B is not small then the d.c. input resistance (R_{in}) between base and ground should be included. This is shown in Fig. 3.35. It can be shown that

$$R_{\text{in}} = \beta_{\text{dc}} R_E \tag{3.22}$$

and generally if $R_{\text{in}} = 10R_2$ it can be ignored. From Fig. 3.34 if the base current is very small then $I_C = I_E$, hence using Kirchhoff's voltage law through the transistor branch gives

$$V_{\text{CC}} = V_{\text{CE}} + I_C R_C + I_C R_E \tag{3.23}$$

When a load is connected to an a.c. coupled amplifier the output voltage and the voltage across the collector–emitter junction change.

The equivalent voltage divider network circuit is shown in Fig. 3.36. Note that R_L and R_C form a voltage divider network between V_{CC} and ground so

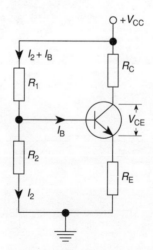

Figure 3.34

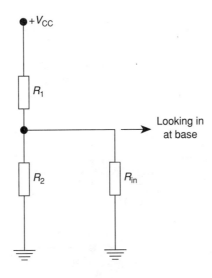

Figure 3.35

that the following expression is valid:

$$V_L = \left(\frac{R_L}{R_L + R_C}\right)(V_{CC} - I_C R_C) \qquad (3.24)$$

Also

$$V_L = V_{CE} + I_C R_E = \left(\frac{R_L}{R_L + R_C}\right)(V_{CC} - I_C R_C) \qquad (3.25)$$

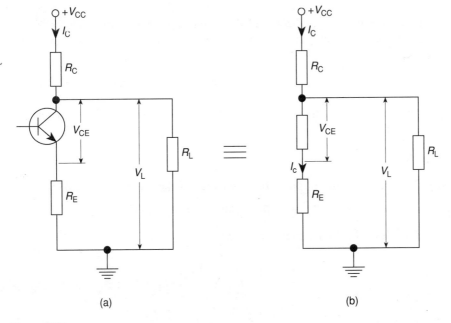

(a)   (b)

Figure 3.36

Example 3.16

The temperature of a liquid in a processing plant is sensed by a thermistor as shown in Fig. 3.37. The thermistor is connected to a d.c. amplifier and the output of this goes to an analogue to digital converter (ADC) which converts the voltage to a digital signal which is then fed to a processor.

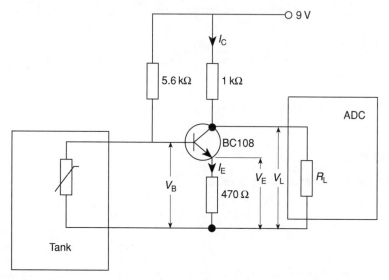

Figure 3.37

The BC108 has a β_{dc} of 200 and the input resistance of the ADC is 50 kΩ. If the response characteristic for the thermistor is shown in Fig. 3.38, determine:

(a) the output voltage (V_L) for the following temperatures: 46, 48 and 51 °C;
(b) whether the transistor goes into saturation or cut-off within these operating temperatures.

Solution

(a) For a temperature of 46 °C the resistance of the thermistor is 2.4 kΩ:

$$R_{in} = \beta_{dc} R_E = 200 \times 470 = 94 \text{ k}\Omega$$

Therefore, $R_{in} \gg R_2$ so it can be ignored.

$$\therefore V_B = \left(\frac{R_2}{R_1 + R_2} \right) V_{CC}$$

$$= \left(\frac{2.4}{5.6 + 2.4} \right) 9 = 2.7 \text{ V}$$

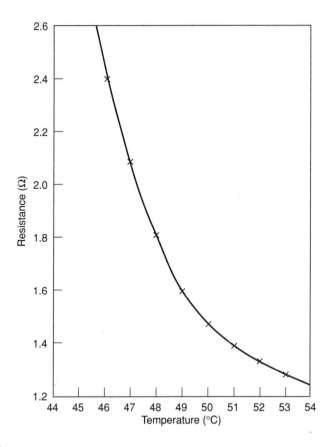

Figure 3.38

Also

$$V_B = V_E + V_{BE}$$

$$\therefore \ V_E = V_B - V_{BE} = 2.7 - 0.7 = 2 \, \text{V}$$

$$\therefore \ I_E = I_C = \frac{V_E}{R_E} = \frac{2}{470} = 4.26 \, \text{mA}$$

$$\therefore \ V_L = \left(\frac{R_L}{R_L + R_C} \right) (V_{CC} - I_C R_C)$$

$$= \left(\frac{50}{51} \right) \left(9 - \frac{4.26}{10^3} \times 10^3 \right) = 4.64 \, \text{V}$$

For a temperature of 48 °C the resistance is 1.8 kΩ.
Once again R_{in} can be ignored and

$$V_B = \left(\frac{1.8}{5.6 + 1.8} \right) 9 = 2.2 \, \text{V}$$

$$\therefore \ V_E = 2.2 - 0.7 = 1.5 \, \text{V}$$

$$I_C = \frac{V_E}{R_E} = \frac{1.5}{470} = 3.19 \, \text{mA}$$

$$\therefore \ V_L = \left(\frac{50}{51}\right)(9 - 3.19) = 5.7 \, \text{V}$$

When temperature is $51 \, °\text{C}$ the resistance of $R_2 = 1.4 \, \text{k}\Omega$. If R_{in} is ignored

$$V_B = \left(\frac{1.4}{5.6 + 1.4}\right) 9 = 1.8 \, \text{V}$$

$$V_E = 1.8 - 0.7 = 1.1 \, \text{V}$$

$$I_C = \frac{V_E}{R_E} = \frac{1.1}{470} = 2.3 \, \text{mA}$$

$$\therefore \ V_L = \left(\frac{50}{51}\right)(9 - 2.3) = 6.57 \, \text{V}$$

(b) In each case

$$V_L = V_{CE} + V_E$$

$$\therefore \ V_{CE} = V_L - V_E$$

and for $46 \, °\text{C}$

$$V_{CE} = 4.64 - 2 = 2.64 \, \text{V}$$

for $48 \, °\text{C}$

$$V_{CE} = 5.7 - 1.5 = 4.2 \, \text{V}$$

for $51 \, °\text{C}$

$$V_{CE} = 6.57 - 1.1 = 5.47 \, \text{V}$$

Hence at no time is the transistor saturated or cut off, i.e. it functions within its linear region.

Example 3.17

A d.c. amplifier has to be designed to the following specification under unloaded conditions:

$$\text{d.c. current gain } (\beta_{\text{dc}}) = 20$$

$$I_B \ (\text{quiescent base current}) = 50 \, \mu\text{A}$$

$$I_C \ (\text{quiescent collector current}) = 1 \, \text{mA}$$

$$V_E \ (\text{emitter voltage}) = 1 \, \text{V}$$

$$\text{Voltage between collector and ground} = 6.5 \, \text{V}$$

$$V_{CC} \ (\text{the supply voltage}) = 12 \, \text{V}$$

Calculate all the components.

Solution

The base current is small, hence

$$I_C = I_E = 1 \, \text{mA}$$

$$R_E = \frac{V_E}{I_E} = \frac{1 \times 10^3}{1} = 1 \, \text{k}\Omega$$

$$V_{CE} = V_{CC} - I_E(R_C + R_E)$$

$$5.5 = 12 - \frac{(R_C + 10^3)}{10^3}$$

$$5.5 \times 10^3 = 12 \times 10^3 - (R_C + 10^3)$$

$$\therefore R_C = 5.5 \, \text{k}\Omega$$

$$R_{in} = \beta_{dc} R_E = 20 \times 10^3 = 20 \, \text{k}\Omega$$

As $R_{in} = 10R_2$, then

$$R_2 = 2 \, \text{k}\Omega$$

Now

$$V_B = V_{BE} + V_E = 0.7 + 1 = 1.7 \, \text{V}$$

$$\therefore I_2 = \frac{1.7 \times 10^6}{2 \times 10^3} = 850 \, \mu\text{A}$$

Finally, by voltage division (Fig. 3.39),

$$V_B = \left(\frac{R_2}{R_1 + R_2}\right) V_{CC}$$

$$\therefore R_1 = \frac{R_2(V_{CC} - V_B)}{V_B}$$

$$= \frac{2(12 - 1.7)}{1.7} = 12.1 \, \text{k}\Omega$$

3.17 Operational amplifier

Up to this point devices which are individually packaged have been considered. These are discrete components. This section deals with linear integrated circuits which consist of many components, including transistors and diodes fabricated on to a single silicon substrate.

A commonly used integrated circuit is the operational amplifier (op-amp) which, because of its versatility, is used in many applications. Because of the range of parameters and applications involved, only the more common aspects of this device will be considered in this text. The ideal op-amp has the following characteristics:

- infinite gain
- infinite bandwidth

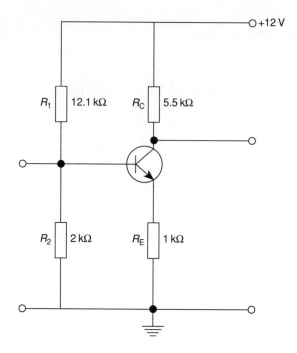

Figure 3.39

- infinite input impedance
- zero output impedance.

The symbol is shown in Fig. 3.40. When the op-amp is connected in this way it is said to be on open loop as no external components are connected between output and input. Such an amplifier, because of its high gain, can amplify unwanted signals such as noise and cause the device to be saturated. Also unwanted oscillations can occur and cause instability and lower the frequency response.

In order to overcome these problems negative feedback is applied. This involves taking a portion of the output and applying it back out of phase with the input. The gain of the amplifier is then much less than the open loop gain and is now called the closed loop gain.

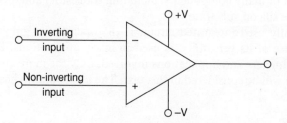

Figure 3.40

3.18 Op-amp configurations

The most common configurations used in control and monitoring systems are:

- the non-inverting amp
- the inverting amp
- the buffer or voltage follower.

Non-inverting amplifier

As can be seen (Fig. 3.41), the input (V_i) is fed to the non-inverting terminals while the feedback is applied to the inverting terminal via a voltage divider network:

$$V_f = \left(\frac{R_i}{R_f + R_i}\right) V_o \qquad (3.26)$$

The gain of the non-inverting amplifier is given as

$$A_c = 1 + \frac{R_f}{R_i} \qquad (3.27)$$

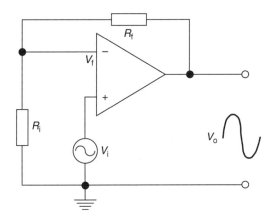

Figure 3.41

Inverting amplifier

This time (Fig. 3.42) the input voltage is fed in at the inverting terminal while the non-inverting terminal is grounded. This is a form of current feedback. Note the gain in this case is

$$A_c = -\frac{R_f}{R_i} \qquad (3.28)$$

Buffer amplifier

If R_f is short-circuited and R_i is open circuited then the gain of the non-inverting

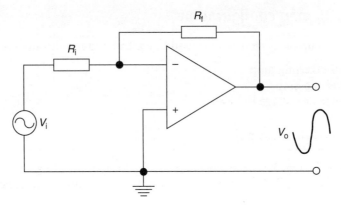

Figure 3.42

amplifier is

$$A_c = 1 + \frac{0}{\alpha}$$

$$\therefore \ A_c = 1$$

The circuit is then configured as in Fig. 3.43. The most important features of the buffer amplifier are its very high input impedance and its very low output impedance. This feature is ideal for interfacing high impedance sources to low impedance sources.

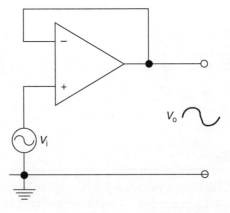

Figure 3.43

Example 3.18

The non-inverting amplifier shown in Fig. 3.44 is required to have a closed loop gain (A_c) of 20. Determine the values of R_f and R_i.

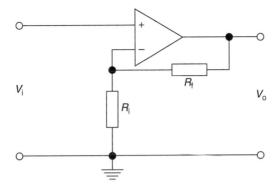

Figure 3.44

Solution
Since the gain has to be 20, select $R_f = 10\,k\Omega$

$$\therefore\ A_c = 1 + \frac{R_f}{R_i}$$

$$20 = 1 + \frac{10}{R_c}$$

$$\therefore\ R_i = 526\,\Omega$$

Select a $1\,k\Omega$ potentiometer.

Example 3.19

An inverting amplifier is used to detect the purity of photographic filters by using an ORP12 photocell. The output is then passed to an analyser for processing. If the light source produces the full visible range from 400 to 700 nm determine:

(a) the voltage gain required and the value to which R_f must be set if the maximum linear output is 2 V less than the supply;
(b) the op-amp output voltage over the range of wavelengths shown in the photocell responses of Fig. 3.45 if the output from the ORP12 is 0.4 V at 780 nm.

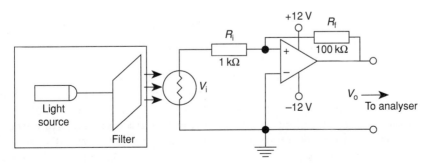

Figure 3.45

Solution

(a) The maximum response occurs at 780 nm and the input at this wavelength is 0.4 V.

$$\therefore A_c = \frac{10}{0.4} = 25$$

$$\therefore R_f \text{ set to } 25 \text{ k}\Omega$$

(b) Selected wavelengths chosen: 500 nm gives

$$V_i = \frac{20}{100} \times 0.4 = 0.08 \text{ V}$$

$$\therefore V_o = 25 \times 0.08 = 2 \text{ V}$$

550 nm gives

$$V_i = \frac{30}{100} \times 0.4 = 0.12 \text{ V}$$

$$\therefore V_o = 25 \times 0.12 = 3 \text{ V}$$

600 nm gives

$$V_i = \frac{60}{100} \times 0.4 = 0.24 \text{ V}$$

$$\therefore V_o = 25 \times 0.24 = 6 \text{ V}$$

700 nm gives

$$V_i = \frac{90}{100} \times 0.4 = 0.36 \text{ V}$$

$$\therefore V_o = 25 \times 0.36 = 9 \text{ V}$$

3.19 Operational amplifier circuits

The previous section investigated op-amp configurations. In this section some of the many specific circuits using op-amps will be discussed.

The comparator

A basic application of the op-amp is its use in determining when an input voltage exceeds a certain threshold level. Figure 3.46 shows a zero-level detector with its input and output waveforms. Because of the large open loop gain of the op-amp it is easily driven into saturation by a small input signal. For example, if the op-amp in Fig. 3.45 had an open loop gain of 200 000, then we would expect an output of ±60 V. However, most op-amps have outputs which swing within 2 or 3 V of the supply voltage, hence the output voltage is ±12 V. Also notice that when the sine wave crosses 0, the amplifier is driven to its maximum positive or negative level.

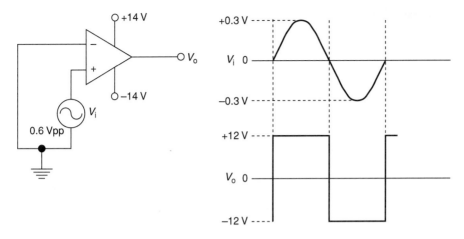

Figure 3.46

A modification of this circuit is shown in Fig. 3.47. In this case a voltage divider has been connected to the inverting input. The voltage applied is called the reference voltage and is calculated as

$$V_{REF} = \left(\frac{R_2}{R_1 + R_2}\right)(+V) = \left(\frac{1}{10+1}\right)(+12) = 1.1\,V$$

This means that each time the input voltage exceeds 1.1 V the output goes to its maximum positive value as shown (output swing has been taken within +3 V of the supply voltage), and if $V_i < V_{REF}$ the output remains at its maximum negative value.

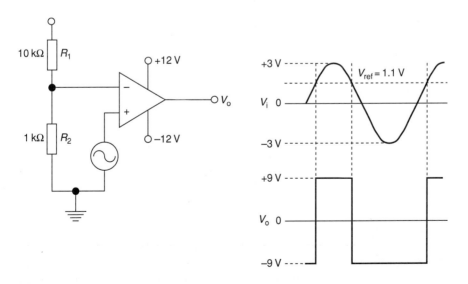

Figure 3.47

Example 3.20

A comparator is connected to a sensor as shown in Fig. 3.48. In this case the zener diode (D_z) produces a reference voltage of 3.2 V. Draw the output waveform from the op-amp if the sensor is ±4 V.

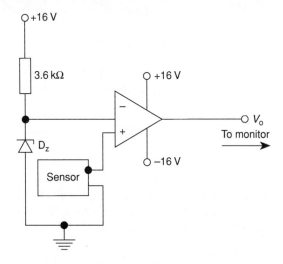

Figure 3.48

Solution
The output waveform is shown in Fig. 3.49.

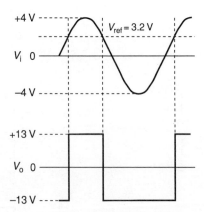

Figure 3.49

Example 3.21

A temperature-sensing device is required to detect when the temperature in a system exceeds 25 °C and then switch the system off. Design a circuit which might accomplish this.

Solution

One method of tackling this problem is to use the knowledge of circuits we have acquired so far. Figure 3.50 shows a suitable circuit. This circuit shows a Wheatstone bridge connected to an op-amp to act as a comparator. The thermistor used is a positive temperature coefficient type (the resistance increases with temperature) and from the data sheets it is found to have a resistance of $3\,k\Omega$ at $25\,°C$.

The potentiometer R_V is set to the value which equals the resistance of the thermistor at $25\,°C$. At temperatures below this value $R_T < R_V$ and this causes the bridge to be unbalanced, thus driving the op-amp to its low saturated output level. This switches T_1 off.

As the temperature increases the thermistor resistance increases and when the temperature reaches $25\,°C$, $R_T = R_V$ and the bridge becomes balanced. The op-amp switches to its high saturated level, turning T_1 on and energizing the relay. In this circuit $R_2 = R_3$, R_4 limits the current to the base of the transistor and T_1 acts as a switch.

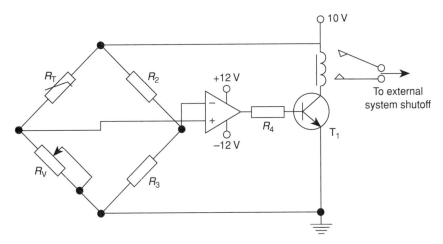

Figure 3.50

Schmitt trigger

When a comparator switches levels, the switching has to be precise and not erratic as may occur if noise or other unwanted signals are present. This erratic performance generally occurs when the comparator is hovering around the reference voltage.

Unwanted signals such as noise causes the effects shown in Fig. 3.51 where noise is superimposed on the input signal (V_i). In order to overcome this problem the comparator is made less sensitive to noise, a method using positive feedback, called hysteresis, is used. In order to appreciate how this works consider Fig. 3.52. R_1 and R_2 form the voltage divider which determines the portion of the output voltage which has to be fed back to the input. There are two triggering points, i.e. the lower (V_{LTP}) and upper (V_{UTP}), and from

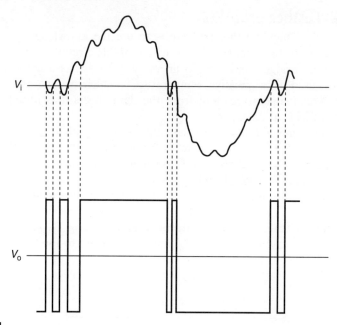

Figure 3.51

Fig. 3.52(b) they are expressed as

$$V_{\text{UTP}} = \left(\frac{R_2}{R_1 + R_2} \right)(+V_{\text{o}}) \tag{3.29}$$

$$V_{\text{LTP}} = \left(\frac{R_2}{R_1 + R_2} \right)(-V_{\text{o}}) \tag{3.30}$$

When the input voltage exceeds V_{UTP}, the output voltage drops to its negative maximum $(-V_{\text{o}})$. The input voltage is now required to fall below V_{LTP} before the device will switch back to its other state. The triggering thus occurs away from the zero crossover points and a small noise voltage has no effect on the output.

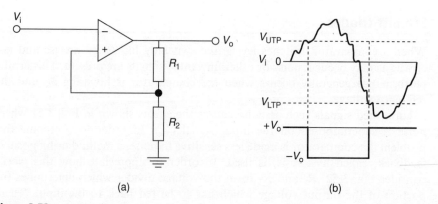

(a)   (b)

Figure 3.52

3.20 Further problems

1. A laser diode emits radiation at 680 nm when using a material with a work function of 0.3. Determine the operational voltage of the laser diode for laser action to take place.
   Answer: 2.14 V

2. A full-wave bridge rectifier has to produce an average current of 1 A in R_L. Determine the following in order to design this unregulated supply which has a secondary voltage of 11 V_{rms}:
   (a) the average current in each diode;
   (b) the minimum PIV of each diode;
   (c) the power dissipated by each diode.
   Answer: 0.5 A; 15.7 V; 25 mW

3. A 50 Hz full-wave bridge rectifier supply has a peak secondary voltage of 18 V at the output of its transformer. If it is supplying a load of 3.6 kΩ at 4.4 mA determine a suitable value of filter capacitor.
   Answer: 3.5 μF

4. A 20 V stabilized supply is required from a 40 V d.c. input voltage. A 20 V zener diode is used having a power rating of 1 W. Determine the value of resistor required to limit the current to the zener.
   Answer: 400 Ω

5. A zener regulator circuit has to be connected to an unregulated supply capable of giving 24 V d.c. The circuit is shown in Fig. 3.53.

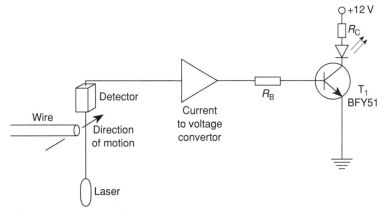

Figure 3.53

6. The zener diode selected as a stabilizer is a 12 V diode with minimum and maximum zener currents of 1 and 50 mA respectively. Determine:
   (a) the minimum and maximum load currents for which the zener will maintain regulation;
   (b) the minimum load resistance that can be used.
   Answer: 0 A, 24.5 A, 490 Ω

7. A full-wave unregulated supply produces a load of $1.5\,\text{k}\Omega$ with a voltage of $200\,\text{V}$ d.c. If the peak-to-peak ripple voltage is $5\,\text{V}$ determine the value of the ripple capacitor.

   Answer: $420\,\mu\text{F}$

8. Two limiter circuits are given in Fig. 3.54. Determine the output waveforms.

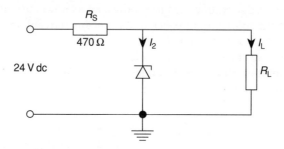

Figure 3.54

9. If the RC time constant in each of the circuits given in Fig. 3.55 is much greater than the period of the input, draw the output waveforms in each case.

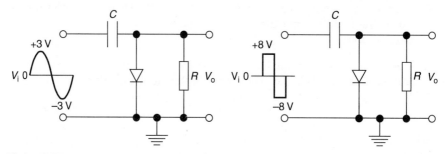

Figure 3.55

10. An optical sensor is used to detect when a wire diameter is outside a certain tolerance. The system is shown in Fig. 3.56. When the output of the I/V converter is $2.5\,\text{V}$ or greater the transistor T_1 saturates. If the β_{dc} of the BFY51 is 60, determine the values of R_B and R_C, assuming that the LED takes $20\,\text{mA}$ and has a forward-biased voltage drop of $1.2\,\text{V}$.

    Answer: $R_C = 1.073\,\text{k}\Omega$; $R_B = 11.25\,\text{k}\Omega$

11. A magnetic switch (S) is used in a security system as shown in the circuit in Fig. 3.57. When the switch is closed T_1 is saturated and $V_{CE(SAT)} = 0.3\,\text{V}$. When the switch is open the base is pulled to ground, T_1 is cut off and approximately $6\,\text{V}$ is applied to T_2 to switch it on. Calculate the values of all the components if the following characteristics are given from the data sheets:

$$\beta_{dc} = 30 \quad V_{CE(SAT)} = 0.3\,\text{V} \quad V_{BE} = 0.6\,\text{V}$$

Answer: $R_{B1} = 100\,\text{k}\Omega$; $R_{B2} = 900\,\text{k}\Omega$; $R_{B3} = 1.2\,\text{k}\Omega$; $R_C = 1.2\,\text{k}\Omega$

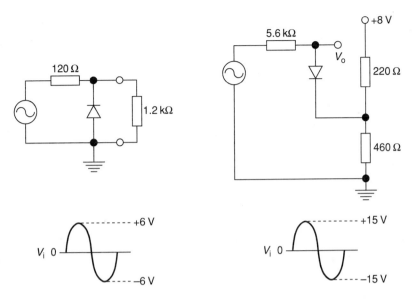

Figure 3.56

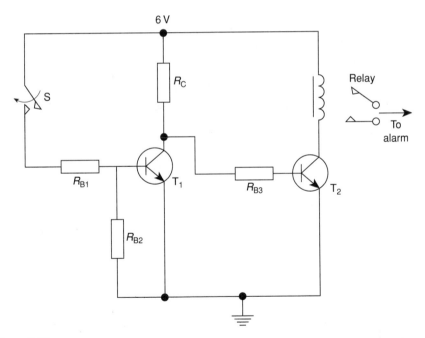

Figure 3.57

12. Design a voltage divider bias circuit in which $V_{CC} = 24\,\text{V}$, the quiescent value of $V_{CE} = 12\,\text{V}$ and the quiescent collector current is $1\,\text{mA}$. A transistor with a $\beta_{dc} = 50$ is used.

Answer: $R_E = 2.4\,\text{k}\Omega$; $R_C = 9.6\,\text{k}\Omega$; $R_1 = 80.9\,\text{k}\Omega$; $R_2 = 120\,\text{k}\Omega$

(Any assumptions made should be stated.)

13. A d.c. coupled amplifier has to be designed to the following specifications under unloaded conditions:

$$\beta_{dc} = 50 \qquad I_C = 1.2\,\text{mA}$$
$$V_E = 0.1V_{CC} \qquad V_{CE} = 7\,\text{V}$$
$$V_{CC} = 15\,\text{V}$$

Calculate all the components and select $R_2 = 15\,\text{k}\Omega$.
Answer: $R_1 = 87.3\,\text{k}\Omega$; $R_C = 5.4\,\text{k}\Omega$; $R_E = 1.25\,\text{k}\Omega$

14. A simple strain gauge circuit is used to drive a small chart recorder motor. The circuit is shown in Fig. 3.58. A cascaded d.c. amplifier is used and the d.c. bias of the first stage sets the d.c. bias of the second. If the chart recorder has an input resistance of $217\,\Omega$, determine:
    (a) the value of all the components assuming the strain gauge (SG) has an unloaded resistance of $120\,\Omega$;
    (b) the current range over which the recorder motor must function if the strain gauge resistance varies between 120 and $122\,\Omega$.
    The transistors are assumed to be identical:

$$V_{CE} = 0.9\,\text{V} \qquad \beta_{dc} = 200 \qquad V_E = \frac{V_{CC}}{10}I_C$$

Also $I_{C1} = 8\,\text{mA}$ and $I_{C2} = 12\,\text{mA}$ when SG $= 120\,\Omega$.
Answer: $R_1 = 698\,\Omega$; $R_{C1} = 1.575\,\text{k}\Omega$; $R_{C2} = 1.033\,\text{k}\Omega$; $R_{E1} = 188\,\Omega$; $R_{E2} = 142\,\Omega$; 8.72–9.9 mA

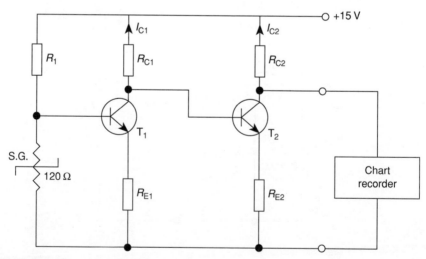

Figure 3.58

15. A tracking system in a CD player (Fig. 3.59) uses two photodiodes to detect when the laser is misaligned relative to the disc tracks. The current output from the photodiode is directly proportional to the incident light intensity, but the processing and control are only calibrated for 20, 40, 60, 80 and $100\,\mu\text{A}$. Determine:

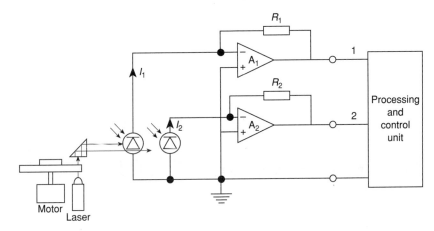

Figure 3.59

 (a) the values of R_1 and R_2 if the maximum rated voltages appearing at inputs 1 and 2 have to be 1 and 2.2 V_{rms} respectively;
 (b) the corresponding input voltages appearing at the processing unit.

16. The circuit in Fig. 3.60 uses a photoconductive cell to switch a power transistor and reset an external circuit. Determine:
 (a) the minimum value of resistance the photoconductive cell must have in order to operate the reset;
 (b) the minimum value of R_S.
 Answer: 61.2 kΩ; 56 kΩ

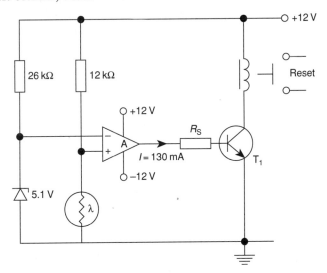

Figure 3.60

17. A sensing circuit has to be designed which will measure the tensile load in a ring. An instrumentation amplifier is required which will give a closed loop voltage gain of 300. Investigate this problem and draw a possible schematic diagram complete with sensing circuit and instrumentation amplifier.

4

Applications of electromagnetic induction

4.1 Introduction

The first magnetic phenomena were associated with natural magnets such as magnetite and lodestone. Industrially based materials used in transformers, motors and magnetic switches are generally, iron, steel, nickel and cobalt. These are known as ferromagnetic materials. However, commercially used materials used in magnetic devices are normally formed from alloys known as ferrous materials. These are artificial magnets which have the advantages of being lighter and also possessing stronger attractive properties (permanent magnets).

A common application of an electromagnet is the recording process on magnetic tape which is shown in Fig. 4.1. The recording head is an electromagnet with an air gap. This application and others will be discussed in this chapter.

4.2 Molecular structure of a magnet

All materials are affected by the presence of a magnetic field, some in different ways. In order to understand this it is necessary to briefly consider the structure and magnetic properties of atoms.

When an electron is moving in its orbital path it constitutes an electric current. This in turn sets up a magnetomotive force (MMF). However, the electrons do not all move in the same direction around the nucleus, so the resultant magnetic effect of the orbital motion of all the electrons can vary from atom to atom.

An MMF may also be produced due to electron spin which can be in either of two directions about its own axis. Two electrons in the same shell and having opposite spins produce MMFs which cancel external to the atom (Fig. 4.2). However, some atoms do not have even numbers of electrons in all their shells and we therefore say that one or more of the spins must be uncompensated.

Modern theory suggests that it is the magnetic moment of the axial spins of the electrons which has the greatest influence on the magnetic properties of materials. All materials whose atoms possess magnetic moments, whatever

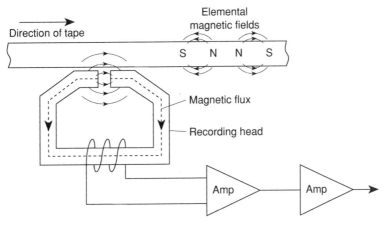

Figure 4.1

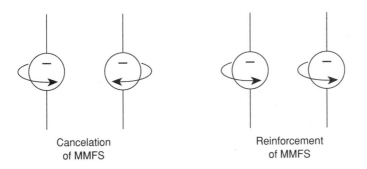

Figure 4.2

their origin, can be separated into two groups, namely those whose atoms have permanent magnetic moments due to uncompensated axial spins and those that do not. Materials which do not possess permanent magnetic moments are called diamagnetic. The remaining group of materials can be further classified according to the interaction between the permanent magnetic moments into paramagnetic, ferromagnetic, antiferromagnetic and ferrimagnetic materials.

In electronic engineering ferromagnetic and ferrimagnetic materials are most common. Both these materials are very similar, but in ferromagnetic materials the magnetic moments tend to line up in parallel (as shown in Fig. 4.3(c)) when placed in a magnetic field.

Neighbouring permanent magnetic moments in a ferromagnetic material are also in parallel and act in opposition, but they are not equal so the resultant magnetic moment may be large. Ferromagnetic materials, usually known as ferrites, are oxides of ferromagnetic materials and exhibit similar effects.

The molecules in most materials are small magnets, but since they are arranged at random no resultant magnetic effect is observed. This is shown in Fig. 4.3(a). In ferromagnetic materials the molecules form a number of closed loops called magnetic domains. There is no resultant magnetic field and the domains remain closed as in Fig. 4.3(b). If the material is subjected to an external magnetic

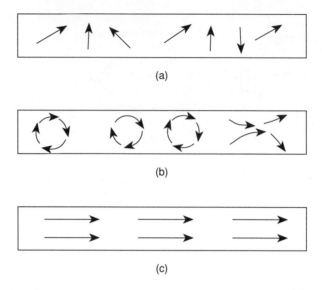

Figure 4.3

field the magnetic domains can be forced open and the molecules align themselves, thus creating a small external magnetic field as in Fig. 4.3(c). If the external field is removed the domains return to their original positions. If this happens the material is known as soft iron. If the molecules retain their position for some time then the material is known as a permanent magnet.

When a bar magnet is formed it has a north and south pole which produces lines of force or magnetic flux as shown in Fig. 4.4. Magnetic flux (Φ) is given in webers (Wb). It is taken as the total magnetism flowing across a given area.

Multiple magnetic systems occur in devices such as stepper motor coils and d.c. motors and these generally obey the simple laws that like poles repel and unlike poles attract. As with an electrostatic field, the concentration of the magnetic field lines is an indication of the strength of the magnetic field. The greater the density of the lines the stronger the magnetic field. This is measured by the amount of magnetic flux passing through a defined area at right angles to the direction of the flux (Φ). The strength of the magnetic flux is known as the magnetic flux density (B) and is given in tesla (T) or webers/m^2:

$$B = \frac{\Phi}{A} \qquad (4.1)$$

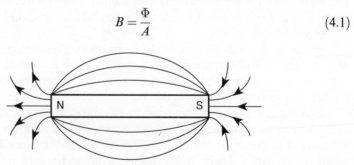

Figure 4.4

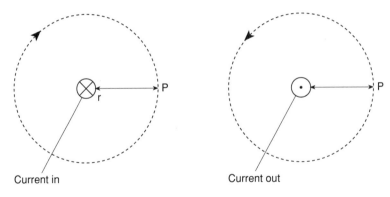

Current in   Current out

Figure 4.5

4.3 Field due to a coil or solenoid

Figure 4.5 shows a cross-section of two long straight conductors carrying a current (I). The direction of the circular field can be determined by the direction of the current. Since the conductor is virtually a single-turn coil, the MMF acting around any of the lines of force is equal to I. Also the length of a line of force passing through a point P in either case, r metres from the centre of the conductor, is $2\pi r$ metres. Hence the magnetizing force at any point around this line of force is

$$H = \frac{I}{2\pi r} \,\text{At}\,\text{m}^{-1} \tag{4.2}$$

The strength of the field round a piece of wire is comparatively weak, but it can be increased by winding it round magnetic material. This is the basis of the electromagnet, the polarity of which is shown in Fig. 4.6(b) by means of the right-hand grip rule.

4.4 Permeability

At any point in a magnetic field the magnetizing force (H) maintains a magnetic flux and produces a certain value of flux density at the point. If H changes, B also changes in proportion. If the magnetic field is in air the ratio B to H is found to be constant. This constant is called the permeability of free space (μ_0). It has the value $4\pi \times 10^{-7}\,\text{H}\,\text{m}^{-1}$. Hence

$$B = \mu_0 H \tag{4.3}$$

If an iron core is inserted into the coil the value of the flux density will increase and the ratio of the flux density produced with the core to the flux density without the core is called the relative permeability (μ_r) of the iron. Hence

$$B = \mu_0 \mu_r H = \mu H \tag{4.4}$$

where μ is the absolute permeability.

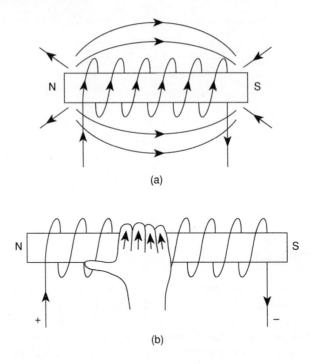

(a)

(b)

Figure 4.6

4.5 Magnetic circuits

Magnetic circuits are used in transformer motors and relays and they have an analogy to the simple d.c. electrical circuit. The close relationship between the parameters of the circuits is shown in Fig. 4.7. Note that the magnetomotive

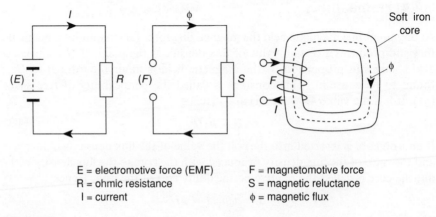

E = electromotive force (EMF)   F = magnetomotive force
R = ohmic resistance   S = magnetic reluctance
 I = current ϕ = magnetic flux

Figure 4.7

force (MMF) is sometimes called the ampere-turns (At) and is expressed as

$$F = NI \qquad (4.5)$$

Also

$$\Phi = \frac{\text{MMF}}{S} \qquad (4.6)$$

The magnetic reluctance is given by

$$S = \frac{1}{\mu A} \qquad (4.7)$$

It seems reasonable to assume that a long magnetic circuit would impede the flux more than a short magnetic path. Similarly the larger the cross-sectional area, the more flux will flow compared to a small cross-sectional area. Finally, the greater the absolute permeability the greater the flux and hence the smaller the reluctance.

As with electrical circuits, magnetic circuits can be either series or parallel and these will now be considered in their various forms.

Series magnetic circuits

As with electrical circuits, magnetic circuits may have different reluctances in series (resistances). This may be due to different materials or air gaps.

Example 4.1

A uniform iron ring used as part of a record head has a cross-sectional area of 450 mm^2 and a mean circumference of 140 mm as shown in Fig. 4.8. An air gap of 1.5 mm is cut in the ring and 1000 turns wound around it. If $\mu_r = 348$ calculate the current required in the coil to produce the required flux of 0.4 Wb in the complete magnetic circuit.

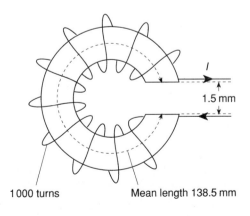

1000 turns   Mean length 138.5 mm

Figure 4.8

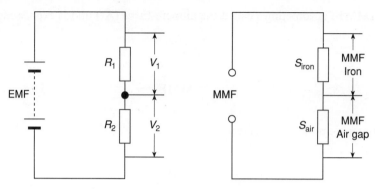

Figure 4.9

Solution

For this first example the analogy to the electrical circuit is given (Fig. 4.9). The length of the iron is equal to the mean circumference minus the air gap (ag):

$$140 - 1.5 = 138.5 \, \text{mm}$$

$$S_i = \frac{l_i}{\mu_0 \mu_r A} = \frac{148 \times 10^{-3}}{348 \times 4\pi \times 10^{-7} \times 450 \times 10^{-6}}$$

$$= 752 \times 10^3 \, \text{At} \, \text{Wb}^{-1}$$

$$S_{\text{ag}} = \frac{l_{\text{ag}}}{\mu_0 A} = \frac{1.5 \times 10^{-3}}{4\pi \times 10^{-7} \times 450 \times 10^{-6}}$$

$$= 2.65 \times 10^6 \, \text{At} \, \text{Wb}^{-1}$$

$$\therefore \ S_T = S_i + S_{\text{ag}} = 752 \times 10^3 \times 2.65 \times 10^6$$

$$= 3.4 \times 10^6 \, \text{At} \, \text{Wb}^{-1}$$

$$\text{MMF} = \Phi S_T = 0.4 \times 10^{-3} \times 3.4 \times 10^6$$

$$= 1.36 \times 10^3 \, \text{At}$$

$$\therefore \ \text{MMF} = NI$$

so

$$I = \frac{\text{MMF}}{N} = \frac{1.36 \times 10^3}{1000} = 1.36 \, \text{A}$$

Magnetizing curves

The reluctance is not often used in practical calculations relating to magnetism. Instead the flux density is calculated and then reference is made to a magnetization curve for the particular material to find H.

It was stated earlier that the ratio B/H for a ferromagnetic material was constant. This is only true over a certain magnetizing range. Figure 4.10

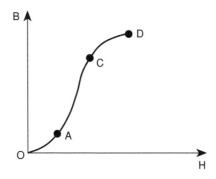

Figure 4.10

shows a magnetizing force applied to a ferromagnetic material which is initially unmagnetized. As an external magnetizing force is applied to the material the molecular domains start to align themselves with the applied magnetizing force. Between O and A this is nonlinear. From A to C the change in this density is proportional to the change in magnetizing force. If H is increased even further, a point is eventually reached at which all the domains are aligned with the applied magnetic field. Any further increase in H produces no further increase in B, and the material is said to be saturated as at D. The linear section AC is generally used in calculations.

Example 4.2

A toroid has to be constructed to produce a magnetic flux of 0.1775 mWb inside the ring (most of the magnetic field is enclosed by the windings in a toroid). The ring dimensions are shown in Fig. 4.11. Determine the current (I) and relative permeability under these conditions.

Solution
The magnetization curve is shown in Fig. 4.12. Diameter of cross-section of toroid = outer radius – inner radius = 11 – 9 = 2 cm. Cross-sectional area of

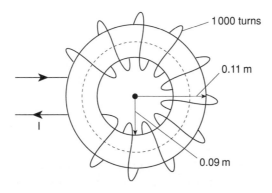

Figure 4.11

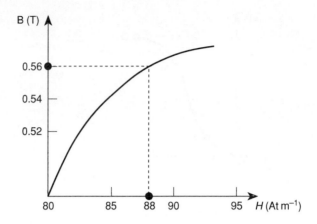

Figure 4.12

toroid is

$$A = \pi \left(\frac{2 \times 10^{-2}}{2} \right)^2 = \pi \times 10^{-4}\,\mathrm{m}^2$$

$$B = \frac{\Phi}{A} = \frac{0.1775 \times 10^{-3}}{\pi \times 10^{-4}} = 0.565\,\mathrm{T}$$

From the magnetizing curve shown in Fig. 4.12

$$H = 88\,\mathrm{At\,m}^{-1} \quad \text{and} \quad B = 0.565\,\mathrm{T}$$

$$\text{Mean radius } r = \frac{\text{outer radius} + \text{inner radius}}{2} = 10\,\mathrm{cm}$$

∴ Mean length of the magnetic circuit $l = 2\pi r = 2\pi \times 10 \times 10^{-2} = 0.2\pi\,\mathrm{m}$. Since MMF $= Hl$, then

$$NI = Hl$$

$$\therefore I = \frac{Hl}{N} = \frac{88 \times 0.2\pi}{1000} = 55.3\,\mathrm{mA}$$

Now

$$\mu = B/H = \mu_0 \mu_\mathrm{r}$$

$$\therefore \mu_\mathrm{r} = \frac{B}{\mu_0 H} = \frac{0.565}{4\pi \times 10^{-7} \times 88} = 5107$$

Example 4.3

A composite series magnetic circuit has to be designed to produce a current of 5 A. The circuit is shown in Fig. 4.13. If the following data are given for the ring calculate the MMF required to produce a magnetic flux of 4 mWb in the ring and also the number of turns.

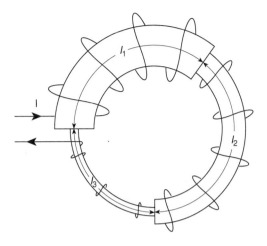

Figure 4.13

Section	μ_r	A	l
1	1	200 mm²	1 mm
2	2000	400 mm²	80 mm
3	800	600 mm²	120 mm

Solution

This type of problem is most conveniently tackled by tabulating the values as shown in Table 4.1.

$$\text{Total MMF} = \frac{10^5}{2\pi} + \frac{10^3}{\pi} + \frac{10^4}{4\pi}$$

$$= \frac{10^3 \times 214}{4\pi} = 17\,020\,\text{At}$$

Since $\text{MMF} = NI$

$$N = \frac{\text{MMF}}{I} = \frac{17\,020}{5} = 3404$$

Parallel magnetic circuits

The parallel magnetic circuit is similar to the parallel electric circuit. The method of approach is shown in the following example.

Example 4.4

A magnetic wrought iron circuit is shown in Fig. 4.14 in which the coil is wound on the centre limb. If the centre limb has a cross-sectional area of 800 mm and each side limb has a cross-sectional area of 400 mm², calculate the MMF required to produce a flux of 1 mWb in the centre limb.

Table 4.1

Section	I	A	Φ	$B = \Phi/A$	μ_r	$H = B/\mu$	MMF
1	1×10^{-3}	200×10^{-6}	4×10^{-3}	$\dfrac{4 \times 10^{-3}}{200 \times 10^{-6}} = 20$	1	$\dfrac{20}{4\pi \times 10^{-7} \times 1} = \dfrac{10^8}{2\pi}$	$\dfrac{10^8}{2\pi} \times 10^{-3}$
2	80×10^{-3}	400×10^{-6}	4×10^{-3}	$\dfrac{4 \times 10^{-3}}{400 \times 10^{-6}} = 10$	2000	$\dfrac{10}{4\pi \times 10^{-7} \times 2000} = \dfrac{10^5}{8\pi}$	$\dfrac{10^5}{8\pi} \times 80 \times 10^{-3}$
3	120×10^{-3}	600×10^{-6}	4×10^{-3}	$\dfrac{4 \times 10^{-3}}{600 \times 10^{-6}} = \dfrac{20}{3}$	800	$\dfrac{20}{3 \times 4\pi \times 10^{-7} \times 800} = \dfrac{10^6}{48\pi}$	$\dfrac{10^6}{48\pi} \times 120 \times 10^{-3}$

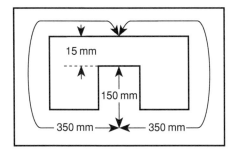

Figure 4.14

Solution

An electrical analogy to the given magnetic circuit is shown in Fig. 4.15. In this case R_a = air gap reluctance, R_b = centre limb reluctance and R_c = outer limb reluctance. The magnetic circuit can be tackled in the same way and the MMF calculated by finding what amounts to the sum of the 'magnetic p.d.s' across each part of the simplified circuit shown in Fig. 4.15(a).

For air gap:

$$\Phi = 1 \times 10^{-3} \, \text{Wb}$$

$$B = \frac{\Phi}{A} = \frac{1 \times 10^{-3}}{800 \times 10^{-6}} = 1.25 \, \text{T}$$

$$H = \frac{B}{\mu_0}$$

$$\text{MMF} = Hl$$

$$= \frac{Bl}{\mu_0} = \frac{1.25 \times 15 \times 10^{-3}}{4\pi \times 10^{-7}} = 1492 \, \text{At}$$

For the centre limb: the B/H curve for wrought iron is given in Fig. 4.16. B in the centre limb is 1.25 T and from Fig. 4.16 this gives

$$H = 750 \, \text{At m}^{-1}$$

$$\therefore \quad \text{MMF} = 750 \times 0.15 = 112.5 \, \text{At}$$

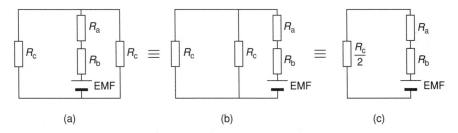

(a)   (b)   (c)

Figure 4.15

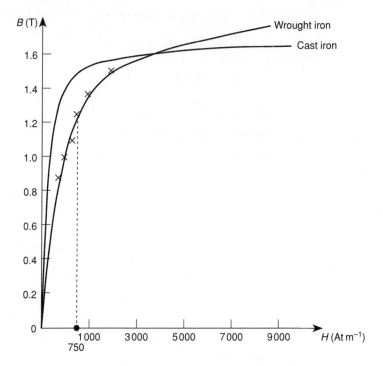

Figure 4.16

For the outer limbs: these limbs are in parallel, thus they may be replaced by one limb of twice the area and the same length as either:

$$\therefore \ B \text{ for the equivalent limb} = \frac{1 \times 10^{-3}}{800 \times 10^{-6}} = 1.25 \,\text{T}$$

$$\text{MMF} = 750 \times 0.35 = 262.5 \,\text{At}$$

$$\therefore \ \text{Total MMF} = 1492 + 112.5 + 262.5 = 1867 \,\text{At}$$

Example 4.5

A transformer core and yoke have the shape shown in Fig. 4.17. The mean lengths of the magnetic flux paths are 200 mm in each outer limb and 80 mm in the centre limb. If the cross-sectional area of all the limbs is 400 mm² calculate the ampere-turns required to set up a magnetic flux of 4π mWb in the centre limb. Assume $\mu_{\rm r} = 1000$.

Solution

With symmetrical parallel magnetic circuits such as this one the technique is to divide it into the greatest possible number of symmetrical sections, and dividing the total flux so that the same proportion flows in each corresponding part of the symmetrical sections.

Figure 4.17 shows how the circuit is effectively divided into four symmetrical sections. Hence using the data given:

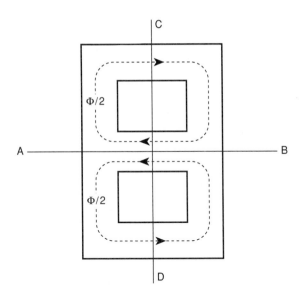

Figure 4.17

Part	Length	A	Φ
$\frac{1}{4}$ centre limb	40 mm	200 mm^2	2π
$\frac{1}{2}$ outer limb	100 mm	400 mm^2	2π

Tabulating the values gives those shown in Table 4.2:

$$\text{Total MMF for half the mean flux path} = 1000 + 1250 = 2250 \, \text{At}$$

$$\therefore \text{Total MMF} = 2 \times 2250 = 4500 \, \text{At}$$

Example 4.6

Figure 4.18 shows a cast-iron magnetic circuit in which a coil of 1500 turns is wound on the centre limb. The cross-sectional area of outer limbs is 1000 mm^2 and of the inner limb 1600 mm^2. Calculate the current required in this coil to produce a flux density of 1 T in the air gap.

Solution
The equivalent electrical circuit is shown in Fig. 4.19. Taking path ADB, for air gap

$$B = 1 \, \text{T}$$

$$H = \frac{1}{4\pi \times 10^{-7}} \, \text{At m}^{-1}$$

$$\text{MMF} = \frac{1 \times 0.12 \times 10^{-2}}{4\pi \times 10^{-7}} = 956 \, \text{At}$$

Table 4.2

Part	l	A	Φ	B	μ_r	H	MMF
$\frac{1}{4}$ centre limb	40 mm	200 mm²	$2\pi \times 10^{-3}$	$\dfrac{2\pi \times 10^{-3}}{200 \times 10^{-6}} = 10\pi$	1000	$\dfrac{10\pi}{4\pi \times 10^{-7} \times 10^3} = \dfrac{10^5}{4}$	$\dfrac{10^5}{4} \times 40 \times 10^{-3} = 10$
$\frac{1}{2}$ outer limb	100 mm	400 mm²	$2\pi \times 10^{-3}$	$\dfrac{2\pi \times 10^{-3}}{400 \times 10^{-6}} = \dfrac{10\pi}{2}$	1000	$\dfrac{10\pi}{4\pi \times 10^{-7} \times 10^3} = \dfrac{10^5}{8}$	$\dfrac{10^5}{4} \times 100 \times 10^{-3} = 12$

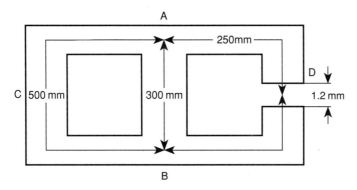

Figure 4.18

For cast iron

$$B = IT$$

From B/H curve for cast iron (Fig. 4.16)

$$H = 200 \, \text{At/m}$$

$$\text{MMF} = 200 \times 0.5 = 100 \, \text{At}$$

$$\text{Total MMF for path ADB} = 1056 \, \text{At}$$

The MMF across the parallel path ACB is the same as this:

$$\therefore \ H \text{ for path ACB} = \frac{1056}{0.5} = 2112 \, \text{At m}^{-1}$$

From B/H curve $B = 1.45 \, \text{T}$

$$\therefore \ \Phi_{\text{ACB}} = \frac{1.4 \times 10}{10^4} = 0.001\,45 \, \text{Wb}$$

and

$$\Phi_{\text{ADB}} = \frac{1 \times 10}{10^4} = 0.001 \, \text{Wb}$$

$$\Phi_{\text{T}} = \Phi_{\text{ACB}} + \Phi_{\text{ADB}}$$

$$\therefore \ \Phi_{\text{T}} = 0.002\,45 \, \text{Wb}$$

$$\therefore \ B \text{ in centre limb} = \frac{0.002\,45}{16 \times 10^{-4}} = \frac{24.8}{16} = 1.55 \, \text{T}$$

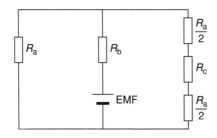

Figure 4.19

From B/H curve $H = 4250\,\mathrm{At\,m}^{-1}$

$$\therefore\ \text{MMF for centre limb} = 4250 \times 0.3 = 1275\,\mathrm{At}$$

$$\text{Total MMF} = 1056 + 1275 = 2331\,\mathrm{At}$$

$$\therefore\ I = \frac{2331}{1500} = 1.55\,\mathrm{A}$$

4.6 Motor effect

It has already been seen that two magnets interact to produce a magnetic field when brought close together. A current-carrying conductor also produces a magnetic field in the form of annular rings and when a current-carrying conductor is placed between two parallel magnetic poles of opposite polarity as shown in Fig. 4.20 an interaction takes place. If the current is entering the page when the field due to the current is clockwise, as shown, the field above the conductor is reinforced while below the conductor the field is weakened. The conductor therefore experiences a force downwards as shown.

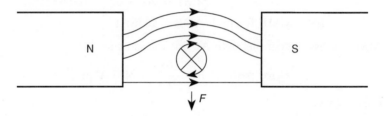

Figure 4.20

If a loop of wire is placed in the magnetic field as in Fig. 4.21 then two opposing forces act on it and a torque is produced which will cause rotation. When the plane of the loop is vertical there is no force acting on it. Only when the loop cuts the magnetic field of the magnets does a force act. Also, if the positive and negative sides of the coil are interchanged, the loop would reverse its direction of motion. Hence some means must be provided to reverse the

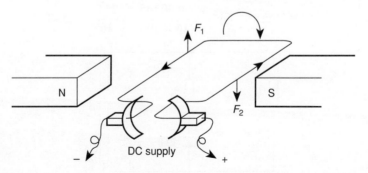

Figure 4.21

direction of current flow in the loop as it crosses the centre position between the two field poles. This is the function of the pair of slip rings shown in Fig. 4.21. The coil is connected to the two rings and this assembly rotates between the two brushes which are connected to the d.c. supply. The coil rotates clockwise and at 90° its momentum causes it to move through the position. The direction of current is now changed in the two limbs and this causes the clockwise rotation of the coils to continue. Hence a continuous torque is achieved.

In practice a motor is constructed of many windings on a former called the armature which may be the rotating member. Furthermore, the permanent magnetic poles are replaced by electromagnets and these are referred to as the field windings. The basic structure is shown in Fig. 4.22.

Finally, a single pair of rings have to be provided for each loop in a practical d.c. motor. This is achieved by means of a commutator which is constructed of individual radial copper segments insulated from each other and from the shaft: the armature coils are connected to these sequences in a certain pattern depending on the type of winding. The insulators between segments are generally made from mica.

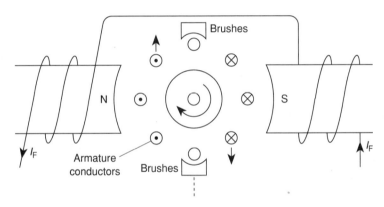

Figure 4.22

Types of windings

Field windings are classified as either shunt or series depending on the intended source of excitation. Where the field and armature windings of a motor are connected in series this is called a series-connected motor, and where the field winding is connected in parallel with the armature this is known as a shunt-connected machine. Both these motor configurations are shown in Fig. 4.23.

DC motor equations

To understand the operating characteristics of the d.c. motor the basic equations will be considered rather than the derivation of these equations.

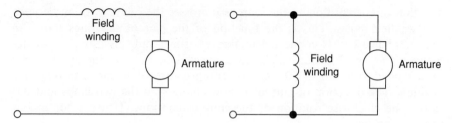

Figure 4.23

Consider Fig. 4.24 and equations (4.8)–(4.13):

$$E \propto I_a N \tag{4.8}$$

$$T \propto I_a^2 \tag{4.9}$$

$$V = E + I_a(R_a + R_f) \tag{4.10}$$

$$E \propto I_f N \tag{4.11}$$

$$T \propto I_f I_a \tag{4.12}$$

$$V = E + I_a R_a \tag{4.13}$$

where E is the induced EMF, V the applied voltage, I_a the armature current, I_f the field current, R_a the armature resistance, R_f the field resistance and N the armature (rev min^{-1}).

The following points should be noted from Fig. 4.24:

- The armature and field d.c. resistances of the series motor are connected in series.
- The armature and field d.c. resistances are connected in parallel in the shunt motor.
- If equation (4.11) is substituted in (4.13) this becomes

$$V \propto I_f N + I_a R_a$$

$$\therefore N \propto \frac{V - I_a R_a}{I_f} \tag{4.14}$$

- This equation indicates that the speed may be increased by decreasing the field current (I_f) or by increasing the armature terminal voltage. In equation (4.14) R_a accounts for all resistance which may appear in the armature. However, any additional resistance in the form of external armature resistance may be used to reduce motor speed.
- A similar reasoning can be applied to the series motor. Substituting equation (4.8) into (4.10) gives

$$N \propto \frac{V}{I_a} - (R_a + R_f) \tag{4.15}$$

- All these equations assume the armature magnetic circuit is unsaturated and that the total flux Φ entering the armature is proportional to I_f ($\Phi \propto I_f$) for a shunt motor and proportional to I_a ($\Phi \propto I_a$) for a series motor.

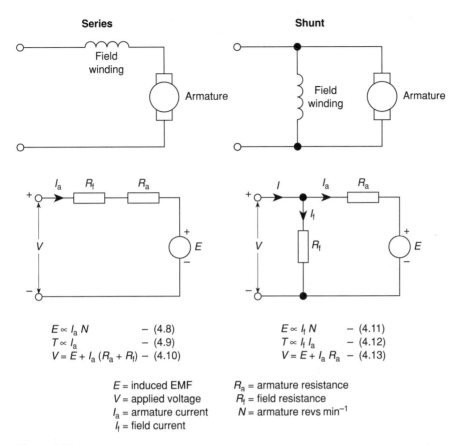

Figure 4.24

Example 4.7

A series motor having a total resistance of $1.2\,\Omega$ takes a current of 24 A from a 250 V supply when running at $1200\,\text{rev min}^{-1}$. If a $2.5\,\Omega$ resistor is placed in series with the motor, calculate the speed at which it will run. Assume the load is being altered to give the same current as before.

Solution

$$N \propto \frac{E}{I_a}$$

$$\therefore\ N \propto \frac{V - I_a(R_a + R_f)}{I_a}$$

$$\frac{N_1}{N_2} = \frac{[V - I_a(1.2)]I_a}{[V - I_a(3.7)]I_a}$$

$$N_2 = 1200\left(\frac{250 - 28.8}{250 - 88.8}\right) = 1647\,\text{rev min}^{-1}$$

Example 4.8

A 220 V shunt motor generally runs at 800 rev min^{-1}. It has an armature resistance of 0.5 Ω and takes an armature current of 25 A. If the motor is required to run at 500 rev m^{-1}, determine the resistance which should be placed in series with the armature. If the load has then to be altered to reduce the armature current to 15 A, what will be the speed of the machine?

Solution
Note there is no information given concerning a change in field current, and since

$$E \propto I_f N$$

then

$$\frac{E_1}{E_2} = \frac{N_1}{N_2}$$

$$\therefore \; E_2 = E_1 \frac{N_2}{N_1} = (220 - 0.5 \times 25)\frac{500}{800} = 130 \, \text{V}$$

Also

$$E_2 = V - I_a(R_a + R_s) \quad (\text{where } R_s = \text{inserted resistance})$$

$$\therefore \; E_2 = V - I_a(0.5 + R_s)$$

The shunt motor characteristic of speed against armature current is very flat, thus I_a in the second case can be taken as 25 A:

$$\therefore \; 130 = 220 - 25(0.5 + R_s)$$

$$\therefore \; R_s = \frac{220 - 130}{25} - 0.5 = 31 \, \Omega$$

The final speed of the motor is

$$E_3 = 220 - 15 \times (3.1 + 0.5) = 166 \, \text{V}$$

$$\frac{E_3}{E_1} = \frac{N_3}{N_1}$$

$$\therefore \; N_3 = \frac{N_1 E_3}{E_1} = \frac{800 \times 166}{220}$$

$$\therefore \; N_3 = 604 \, \text{rev min}^{-1}$$

Example 4.9

A d.c. series motor is connected to a 240 V supply and runs at 50 rev min^{-1} taking a current of 30 A. If the resistance of the motor is 0.3 Ω and it is required to reduce the speed to 600 rev min^{-1}, calculate the value of resistance which must be inserted in series with the motor, the load torque then being half its previous value. Assume flux is proportional to field current.

Solution

$$T \propto I_a^2$$

$$\therefore \frac{T_1}{T_2} = \frac{30^2}{I_{a2}^2}$$

but

$$T_2 = \frac{T_1}{2}$$

$$\therefore I_{a2} = \frac{30}{\sqrt{2}}$$

Also

$$N \propto \frac{E}{I_a}$$

$$\therefore N \propto \frac{V - I_a(R_a + R_f)}{I_a}$$

$$\therefore \frac{N_1}{N_2} = \frac{V - 30 \times 0.3}{V - (30/\sqrt{2})(0.1 + R_s)} \times \frac{30}{30 \times \sqrt{2}}$$

$$\therefore \frac{850}{600} = \frac{240 - 9}{240 - 21.2(0.3 + R_s)} \times \frac{1}{\sqrt{2}}$$

$$\sqrt{2} \times 850[240 - 21.2(0.3 + R_s)] = 600 \times 231$$

$$240 - 6.36 - 21.2R_s = 115.3$$

$$\therefore R_s = \frac{115.3 + 6.36 - 240}{21.2}$$

$$= \frac{240 - 115.3 - 6.36}{21.2} = 5.6 \, \Omega$$

4.7 Electromagnetic induction

In section 4.6 the motor effect was caused by a current-carrying conductor being immersed in a magnetic field. It is also possible to achieve the reverse effect, i.e. moving a conductor in a magnetic field to produce a current in it.

In Fig. 4.25 if the conductor is pulled upwards a current flows in the direction shown. This is called an induced current and the EMF causing it is called an induced EMF. It can be seen that a d.c. generator could be constructed using this principle, but as these are rarely used except as starters to a.c. generators they will not be discussed in this text. However, the principle of electromagnetic induction is an important one and is used in many applications. The main physical principles are discussed below.

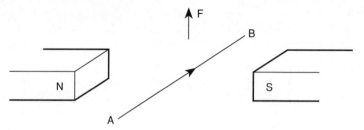

Figure 4.25

Self-inductance

An induced EMF of the type illustrated in Fig. 4.25 depends on the rate of change of flux linking the wire. In Fig. 4.26 an EMF will be induced in the coil, due to the flux linking the coil. Note the magnet may move relative to the coil or vice versa.

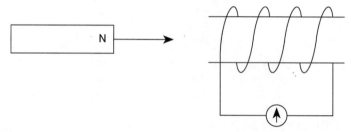

Figure 4.26

In Fig. 4.27 the flux linking the coil before the switch is closed is zero. After the switch has been closed there will be a flux Φ webers linking the coil. This means that an EMF will be induced in the coil, and by Lenz's law its direction will be such as to oppose the change producing it. The coil has induced an EMF in itself and is said to have self-induction. This principle can be summarized in the following expression:

$$e = -N\frac{\Delta\Phi}{\Delta t} \tag{4.16}$$

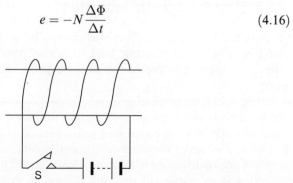

Figure 4.27

where N is the number of turns, $\Delta\Phi/\Delta t$ the rate of change of flux and the negative sign indicates Lenz's law is operating. Since the changing flux is produced by a changing current, it follows that the induced EMF is dependent on the rate of change of current $\Delta I/\Delta t$:

$$e = -L\frac{\Delta I}{\Delta t} \qquad (4.17)$$

where L is the constant called self-inductance measured in henries (H).

Mutual inductance

It has already been shown when a coil is cut by a changing magnetic field an EMF is induced in the coil. Consider two coils A and B shown in Fig. 4.28. When the switch S is closed the current in coil A changes from 0 to a final current of 1 A in a time of t seconds. This results in a change of magnetic flux in coil A from 0 to Φ webers. Part of this changing flux will cut coil B and hence induce an EMF (e_B) in it. The magnitude of the EMF induced in coil B is proportional to the rate of change of magnetic flux which depends on the rate of change of current in coil A. This gives the expression

$$e_B = M\frac{\Delta I_A}{\Delta t} \qquad (4.18)$$

where M is called the mutual inductance. Note that mutual inductance is the ability of one circuit to induce an EMF in another.

If the two coils were a few centimetres apart in air then only a small part of the changing flux produced by the first coil would link the second coil and the circuits would have a low coupling coefficient. However, if the two coils were wound on a soft iron core with the second coil wound very tightly on top of the first, most of the flux produced by the first coil would link the second and the circuits would have a high coupling coefficient. The ideal coupling coefficient is unity and under these circumstances no losses occur.

The mutual inductance can be calculated by

$$M = k_a\sqrt{L_1 L_2} \qquad (4.19)$$

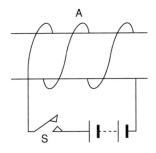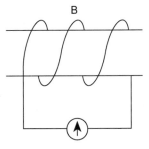

Figure 4.28

4.8 Transformers

Mutual inductance is the basis for the operation of transformers which are used in applications ranging from power supplies and electrical distribution to signal coupling in communications systems.

The transformer consists of two coils placed in close proximity and coupled by their mutual inductance. The coils are wound round magnetic circuits similar to those already mentioned. Instead of solid cores, however, laminated cores are used to reduce eddy currents, while ferrite cores are used for high-frequency applications.

In Fig. 4.29 several points are illustrated. First, the direction of the windings determines the polarity of the voltage across the secondary winding (V_s) with respect to the voltage across the primary (V_p). Phase dots are used to indicate polarities as shown in Fig. 4.29(c). Figure 4.29(a) has the coils wound in the same sense so that the secondary voltage is in phase with the primary voltage. Figure 4.29(b) shows the windings wound in the opposite sense so that the secondary voltage is out of phase with the primary voltage.

Second, in this text only ideal transformers will be considered, i.e. the winding resistance and capacitance as well as the non-ideal core characteristics are all neglected. In order to understand the basic concepts and applications of the transformer this 100 per cent efficiency device is quite adequate. For this reason the d.c. resistance and interwinding capacitance are not shown in Fig. 4.29(c).

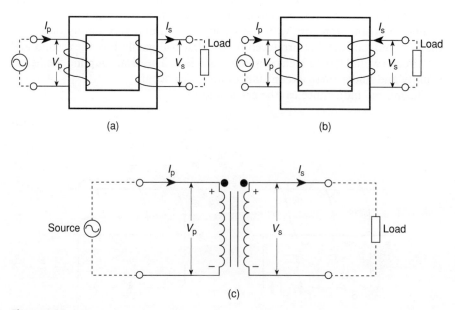

(a)   (b)

(c)

Figure 4.29

Transformation ratio

An important parameter of a transformer is its turns ratio N_s/N_p which is related to the primary and secondary voltages as follows:

$$\frac{N_s}{N_p} = \frac{V_s}{V_p} \qquad (4.20)$$

This relationship can be developed further if the transformer is treated as ideal. In this case all the power is transferred from the primary to the secondary when the secondary is loaded so that

$$P_p = P_s$$

$$I_p V_p = I_s V_s \qquad (4.21)$$

$$\therefore \frac{I_p}{I_s} = \frac{V_s}{V_p}$$

If (4.20) and (4.21) are combined we have

$$\frac{N_s}{N_p} = \frac{V_s}{V_p} = \frac{I_p}{I_s} \qquad (4.22)$$

This expression shows that the voltage and current in the secondary may be increased or decreased, but not both at the same time. Hence if the voltage is stepped up in the secondary the current is stepped down.

Reflected load

From the point of view of primary circuits, a load connected across the secondary of a transformer appears to have a resistance that is not necessarily equal to the actual resistance of the load. We say the actual load is 'reflected' into the primary but altered by the turns ratio. This is now the load that the primary effectively sees.

Assuming 100 per cent efficiency

$$I_p^2 R_p = I_s^2 R_s = I_s^2 R_L$$

$$\therefore \frac{I_p^2}{I_s^2} = \frac{R_L}{R_p}$$

$$\therefore \frac{N_s^2}{N_p^2} = \frac{R_L}{R_p} \qquad (4.23)$$

$$\therefore R_p = \left(\frac{N_p^2}{N_s^2}\right) R_L$$

4.9 Applications of transformers

The main applications of transformers are:

• power supplies

- load matching
- isolation
- autotransformers.

Unregulated power supplies

Power supplies have already been mentioned in this chapter, and at that stage the transformer was used as the step-down device for low-voltage rectified supplies.

Example 4.10

A full-wave unregulated supply has to provide a load of $3\,k\Omega$ at $300\,V$ d.c. Determine:

(a) the r.m.s. voltage of the transformer required;
(b) the turns ratio of the transformer if the peak primary voltage is $650\,V$;
(c) the power rating of the transformer.

Solution

(a)
$$V_{rms} = \frac{V_L}{0.9} = \text{(for full wave)} \ \frac{300}{0.9} = 333\,V$$

(b)
$$\frac{N_s}{N_p} = \frac{V_s}{V_p}$$

$$V_p = \frac{650}{1.414} = 460\,V$$

$$\therefore \ \text{Turns ratio} = \frac{333}{460} = 0.724$$

(c)
$$P_{ac} = I_{rms} \times V_{rms} = \frac{I_{pk}}{1.414} \times \frac{V_{pk}}{1.414}$$

$$I_{pk} = \frac{I_L}{0.636} \quad \text{(for full wave)}$$

and

$$I_L = \frac{V_L}{R_L} = \frac{300}{3000} = 0.1\,A$$

$$\therefore \ I_{pk} = \frac{0.1}{0.636} = 0.157$$

$$\therefore \ P_{ac} = \frac{0.157}{1.414} \times 333 = 36\,W$$

Example 4.11

A full-wave unregulated supply uses a transformer with several secondary windings connected as shown in Fig. 4.30. If the voltage (V_L) across the $15\,\Omega$ load has to be $24\,V$, determine:

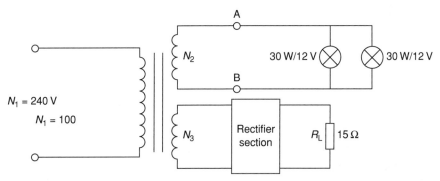

Figure 4.30

(a) N_2 and N_3;
(b) the power rating of the transformer.

Solution

(a) Taking the N_2 circuit, each lamp is rated at $12\,\text{V}$ so $V_2 = 12\,\text{V}$:

$$\therefore \frac{N_2}{N_1} = \frac{V_2}{V_1}$$

$$\therefore N_2 = \frac{V_2}{V_1}N_1 = \frac{12 \times 100}{240} = 5$$

Taking the N_3 circuit

$$V_L = 0.9\,V_{rms}$$

$$\therefore V_{rms} = \frac{24}{0.9} = 267\,\text{V}$$

$$\therefore \frac{N_3}{N_1} = \frac{V_3}{V_1}$$

$$N_3 = \left(\frac{V_3}{V_1}\right)N_1 = \left(\frac{26.7}{240}\right)100 = 11.125$$

(b) In order to find the rating of the transformer the r.m.s. values of the secondary currents and voltages have to be calculated. The r.m.s. current taken by both lamps is

$$I_{rms} = \frac{30}{12\,\text{V}} \times 2 = 5\,\text{A}$$

and the r.m.s. current provided by N_3 is

$$I_{rms} = \frac{I_L}{0.9}$$

$$I_L = \frac{V_L}{R_L} = \frac{24}{15} = 1.6\,\text{A}$$

$$\therefore I_{rms} = \frac{1.6}{0.9} = 1.78\,\text{A}$$

∴ Power taken by N_2 winding is

$$12 \times 5 = 60 \text{ VA}$$

and power taken by N_3 winding is

$$26.7 \times 1.78 = 47.5 \text{ VA}$$

∴ Secondary must have a total rating of 107.5 VA.

Load matching

Frequently it is required to connect a high or low resistance source to a low or high resistance load. If this was carried out there would be a series mismatch in resistances and maximum power would not be transferred. In order to overcome this a matching transformer is used. This is shown in two applications given below.

Example 4.12

In each of the applications shown in Fig. 4.31 determine the turns ratio of each transformer to provide maximum power transfer.

Solution

(a) $\dfrac{N_s}{N_p} = \sqrt{\dfrac{300}{75}} = 2:1$

(b) $\dfrac{N_p}{N_s} = \sqrt{\dfrac{125}{5}} = 5:1$

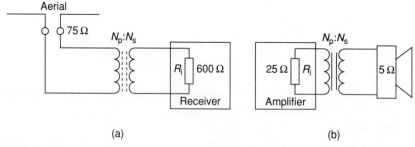

(a)   (b)

Figure 4.31

Isolation

The principle of isolation can be used to isolate a d.c. component or else reduce shock hazards. It should be appreciated by now that an a.c. voltage causes flux linkage in a transformer, i.e. a changing current in the primary is necessary to induce a voltage in the secondary. Hence the transformer isolates the secondary from any d.c. voltages.

This also has implications where a high d.c. voltage may cause a shock hazard if connected to ground or reference. When an isolation transformer is used the secondary circuit is said to be floating because it will no longer be referenced to ground.

Autotransformers

Unlike conventional transformers an autotransformer has one winding which serves as both the primary and secondary. This is achieved by tapping at specific points to achieve the required turns ratio. Autotransformers do not have electrical isolation between the primary and secondary and they are generally smaller and lighter than conventional transformers. These transformers are mainly used for interconnecting systems which operate at approximately the same voltage and for starting up cage-type induction motors.

Example 4.13

An autotransformer has to be used to change a 240 V source voltage to a load voltage of 180 V as shown in Fig. 4.32. Determine the input and output power (kVA) if the transformer is assumed to be ideal and show what is the actual kVA requirement.

Solution
The load current is given as

$$I_3 = \frac{V_3}{R_L} = \frac{180}{12} = 15\,\text{A}$$

$$P_{in} = V_1 I_1 \quad \text{and} \quad P_{out} = V_3 I_3$$

$$P_{in} = P_{out} \quad \text{(ideal)}$$

$$\therefore \ V_1 I_1 = V_3 I_3$$

$$\therefore \ I_1 = \frac{V_3 I_3}{V_1} = \frac{180 \times 15}{240} = 11.25\,\text{A}$$

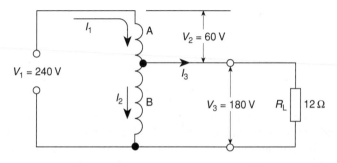

Figure 4.32

Applying Kirchhoff's current law at the tap:

$$I_1 = I_2 + I_3$$

$$\therefore\ I_2 = I_1 - I_3 = 11.25 - 15 = 3.75\,\text{A}$$

The input and output power are

$$P_{\text{in}} = P_{\text{out}} = V_3 I_3 = 180 \times 15 = 2.7\,\text{kVA}$$

The power in winding A is

$$P_A = V_2 I_1 = 60 \times 11.25 = 0.675\,\text{kVA}$$

The power in winding B is

$$P_B = V_3 I_2 = 180 \times 3.75 = 0.675\,\text{kVA}$$

Hence the power rating requirement for windings A and B is less than the power delivered to the $12\,\Omega$ load.

4.10 Further problems

1. A transformer core is made from steel with a relative permeability of 1200 and has an effective magnetic length of 20 cm. A coil of 800 turns is wound round the core. What current is required in the coil to produce a magnetic flux density of 0.8 T?
   Answer: 133 mA

2. A coil is wound uniformly with 300 turns over a steel ring of relative permeability 900 having a mean circumference of 400 mm and a cross-sectional area of 500 mm^2. If the coil has a d.c. resistance of $8\,\Omega$ and is connected across a 20 V d.c. supply, calculate:
   (a) the coil MMF
   (b) the magnetic field strength
   (c) the total flux
   (d) the reluctance of the ring.
   Answer: 750 At; 1875 At m^{-1}; 1.06 mWb; 707 kAt Wb^{-1}

3. An iron ring has a uniform cross-sectional area of $3 \times 10^{-4}\,\text{m}^2$ and a mean circumference of 0.25 m. A coil of 200 turns is wound uniformly on the ring and a current of 0.8 A is passed through it, $\mu_r = 800$. Calculate:
   (a) the MMF
   (b) the magnetic field strength
   (c) the reluctance
   (d) the magnetic flux in the core.
   Answer: 160 At; 640 At m^{-1}; 829 kA Wb^{-1}; 193 µWb

4. A cast steel electromagnet of horseshoe shape has a magnetic length including the armature of 800 mm and a cross-sectional area of 20 cm^2. Assuming the lengths of each of the two air gaps to be 1 mm, find the ampere-turns required to produce a flux density of 1.1 T.

The magnetic data for cast steel are:

H (At m^{-1})	1000	1100	1200	1300
B (T)	0.975	1.055	1.115	1.16

Answer: 1817.5 At

5. A magnetic circuit consists of two portions of stalloy iron having the following dimensions:

	Length	Cross-sectional area
A	40.64 cm	212.9 cm^2
B	30.48 cm	258 cm^2

If section A has an air gap 0.254 cm wide, find the number of ampere-turns required to produce 0.016 Wb in the gap. The B/H curve for stalloy iron gives:

H	80	160	240	320	480	640
B	0.4	0.63	0.8	0.9	0.98	1.8

Answer: 1643.7 At

6. A transformer core and yoke have the shape shown in Fig. 4.33. If the cross-sectional area of all the limbs is 4 cm^2, $\mu_r = 1000$ and a magnetic flux of 12.57 mWb has to be produced in the gap, calculate the MMF.

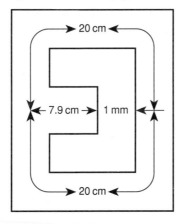

Figure 4.33

Answer: 29 500 At

7. A 250 V shunt motor has an armature resistance of 0.5 Ω and runs at 1200 rev min^{-1} when the armature current is 80 A. If the torque remains unchanged find the speed and the armature current when the field is strengthened by 25 per cent.
Answer: 998 rev min^{-1}; 64 A

8. A 240 V d.c. shunt motor runs at $800 \, \text{rev min}^{-1}$ with no extra resistance in the field or armature circuits, on no-load. Determine the resistance to be placed in series with the field so that the motor may run at $950 \, \text{rev min}^{-1}$ when taking an armature current of 20 A. The field resistance is $160 \, \Omega$ and the armature resistance $0.4 \, \Omega$. It may be assumed that the flux is proportional to the field current.
Answer: $37 \, \Omega$

9. A 440 V series motor has a regulating resistance R connected in series. The motor resistance is $0.3 \, \Omega$. When $R = 0$ and the current $I = 20 \, \text{A}$ the motor runs at $1200 \, \text{rev min}^{-1}$. Find the speed when $R = 3 \, \Omega$ and $I = 15 \, \text{A}$ given that the flux with $I = 15 \, \text{A}$ is 80 per cent of that with $I = 20 \, \text{A}$.
Answer: $1350 \, \text{rev min}^{-1}$

10. A 500 V d.c. shunt motor has a speed of $1000 \, \text{rev min}^{-1}$ when running on a light load. Its armature and brush resistance totals $2 \, \Omega$. In order to reduce the speed of the armature the voltage (V) is tapped off a $50 \, \Omega$ resistor across the supply. If the tapping point is $25 \, \Omega$, at what speed will the motor run when:
(a) the supply current is 15 A;
(b) the armature current is 15 A?
Answer: $212 \, \text{rev min}^{-1}$; $66 \, \text{rev min}^{-1}$

11. A 50 Hz single-phase transformer has square cores of 20 cm side. The permissible maximum flux density is 1 T. Calculate the required number of turns per limb on the high and low voltage sides for a 3000/220 voltage ratio.
Answer: 191:14

12. An 11 500/2300 transformer is rated at 100 kVA as a two-winding transformer. If the two windings are connected in series to form an autotransformer what are the possible voltage ratios and outputs?
Answer: 13.8/11.5 kVA; 600 kVA or 13.8/2.3 kVA; 120 kVA

13. In the circuit in Fig. 4.34 an average power transfer of 100 mW has to be applied to the load R_L. Find:
(a) the r.m.s. value of V_s
(b) the efficiency of this transfer.
Answer: 2.52 V; 62.5 per cent

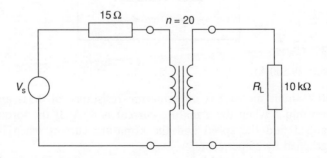

Figure 4.34

14. It is required to drive a 3.5 V a.c. motor using the network shown in Fig. 4.35. Determine the value of R necessary for satisfactory operation.

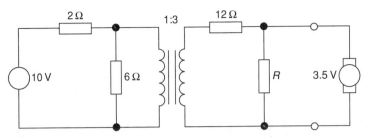

Figure 4.35

Answer: 2.2 Ω

15. A full-wave unregulated supply uses a transformer with several windings connected as shown in Fig. 4.30. The load (R_L) is 220 Ω and the voltage (V_L) across it has to be 12 V. The winding N_2 has three parallel loads having ratings of 10 W/6 V, 5 W/6 V and 14 W/6 V. If the supply is 220 V and $N_1 = 60$, determine:
    (a) N_2 and N_3
    (b) the power rating of the transformer.
    Answer: $N_2 = 1.6$; $N_3 = 3.6$; 29.36 VA

5

Applications of electromagnetic waves

5.1 Introduction

The block diagrams in Fig. 5.1 show the various processes which are involved in two communication systems. Many questions have to be answered before such systems are functional. How does the modulator mix signals? What filter design is required? How is the output connected to the aerial system? These problems may be tackled once a knowledge of electromagnetic waves is acquired.

A wave is a means of transferring energy from one point to another without there being any transfer of matter between these points. In electrical and electronic engineering we generally consider electromagnetic waves which have practical implications in aerial technology, transmission lines, propagation methods and electronic measurement and testing. Note for electromagnetic waves the speed is $3 \times 10^8 \, \text{m s}^{-1}$. Such a wave is shown in Fig. 5.2 where λ is the wavelength (m), V_p the amplitude, V_{pp} the peak-to-peak amplitude, T the period, f the frequency (Hz) and v the velocity (m s^{-1}). Also

$$v = f\lambda \quad \text{and} \quad T = \frac{1}{f}$$

The equation of a progressive wave is given as

$$a = A \sin 2\pi f t = A \sin \omega t \tag{5.1}$$

Note this is the equation for a wave which starts at the origin, i.e. there is no phase shift.

5.2 Phase shift

Generally in electrical work phase shift plays an important part in voltage and current waveforms. This will be explained in more detail later. In Fig. 5.2 a point S is selected which is displaced by x. Since the wave is moving to the right the motion of point S will lag behind that of a point at the origin. In terms of the period this is given as $(x/\lambda)T$. This alters the previous equation

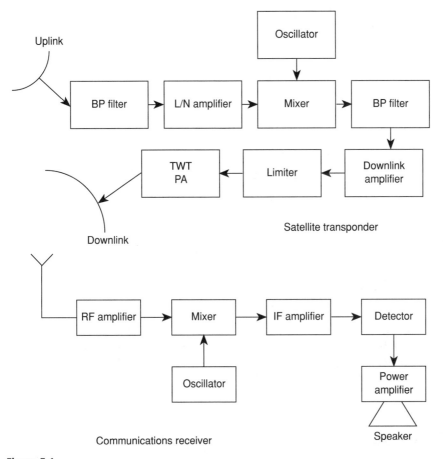

Figure 5.1

as follows:

$$a = A \sin\left\{ 2\pi f \left(t - \frac{xT}{\lambda} \right) \right\}$$

and as $T = 1/f$ then

$$a = A \sin(2\pi ft - kx) \qquad (5.2)$$

where $k = 2\pi/\lambda$. It can be seen that the displacement of S can be expressed as a lagging phase angle in terms of time. If the progressive wave was propagated from right to left the angle would be a leading phase angle for the point S.

5.3 Average and r.m.s. values

The true average of a sine wave is zero, as the positive half-cycle cancels out the negative half-cycle. This is of no practical use and so for electrical purposes the average value refers to the average magnitude. This is of importance when

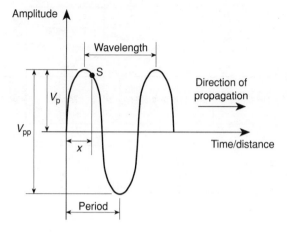

Figure 5.2

considering such waveforms as rectified waveforms where the d.c. or average value is required. The d.c. or average value is generally given as

$$\text{Average value} = 0.637 \times \text{maximum value}$$

Of greater significance than the average value is the effective value. This is defined for alternating currents and voltages as the steady d.c. value that would have the same effect. Generally this is known as the **root mean square** value and is given as

$$\text{r.m.s. value} = 0.707 \times \text{maximum value}$$

5.4 Problems on progressive waves

Example 5.1

A radio station transmits an electric field wave which satisfies the following equation:

$$e = 5 \times 10^{-6} \sin 2\pi \times 92 \times 10^6 t \text{ volts}$$

Determine the wavelength of the station.

Solution
The expression shows that the frequency is 92 MHz, hence the two-dimensional wave equation can be used to find the wavelength:

$$\lambda = \frac{v}{f} = \frac{3 \times 10^8}{92 \times 10^6} 3.26 \text{ m}$$

Example 5.2

A progressive voltage wave has a frequency of 50 Hz. If its peak amplitude is 3 V, determine:

(a) its instantaneous value at an eighth of its period;
(b) its r.m.s. value;
(c) its mean value.

Solution

(a) The period of the wave should be found first:

$$T = \frac{1}{f} = \frac{1}{50} = 0.025 = 20\,\text{ms}$$

Hence

$$\frac{T}{8} = \frac{0.02}{8} = 0.0025\,\text{s} = 2.5\,\text{ms}$$

Since

$$v = V_p \sin 2\pi ft = 3 \sin 2\pi \frac{1}{T} \times \frac{T}{8}$$

$$= 3 \sin \frac{2\pi}{8} = 3 \sin \frac{\pi}{4} = 3 \sin 45° = 2.12\,\text{A}$$

(b) $$V_{\text{rms}} = \frac{V_p}{\sqrt{2}} = 3 \times 0.707 = 2.121\,\text{V}$$

(c) $$V_{\text{mean}} = V_p \times 0.637 = 3 \times 0.637 = 1.911\,\text{V}$$

Example 5.3

(a) Give the general expression for a sinusoidal current of peak value 12 A lagging by $5\pi/6\,\text{rad s}^{-1}$ if the frequency is 100 Hz.
(b) Determine its value at a quarter of the period.

Solution

(a) As the waveform is lagging then

$$i = 12 \sin\left(200\pi ft - \frac{\pi}{6}\right)$$

(b) At a quarter of the period, since $T = 1/100 = 0.01\text{s}$ then

$$i = 12 \sin\left(200\pi \times 0.01 \times 0.25 - \frac{5\pi}{6}\right)$$

$$= 12 \sin\left(\frac{3\pi}{6} - \frac{5\pi}{6}\right) = 12 \sin\left(\frac{-2\pi}{6}\right) = -0.866\,\text{A}$$

Example 5.4

Two 50 Hz currents have peak values of 6 and 2 A (Fig. 5.3). If the 2 A current lags the 6 A current by 24°,

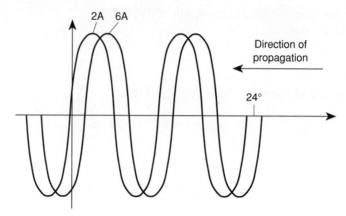

Figure 5.3

(a) give expressions for the two currents;
(b) determine the instantaneous values of the currents after $T/3$.

Solution

(a) The 6 A current may be represented by

$$t = I_p \sin 2\pi ft = 6 \sin 2\pi \times 50t$$

The 2 A current may be represented by

$$t = I_p \sin(2\pi ft - 24°) = 2 \sin\left(2\pi \times 50t - \frac{6\pi}{45}\right)$$

Note all values should be in radians or degrees.

(b) The instantaneous value of the 6 A current is

$$i = I_p \sin 2\pi ft = I_p \sin 2\pi \times \frac{1}{T} \times \frac{T}{3}$$

$$= 6 \sin\frac{2\pi}{3} = 5.196\,\text{A}$$

Example 5.5

A progressive wave of peak amplitude 0.1 V and wavelengths 2 mm travels along a transmission line with a speed of $1\,\text{m}\,\text{s}^{-1}$. Calculate:

(a) the frequency of the wave;
(b) the angular frequency.

Determine:

(c) the progressive wave equation of a particle 1.15 m to the right of the origin;
(d) the transverse voltage displacement of a particle 1.5 m to the right of the origin at $t = 3.25\,\mu\text{s}$.

Solution

(a) $$f = \frac{v}{\lambda} = \frac{1}{2 \times 10^{-3}} = 500 \, \text{Hz}$$

(b) $$\bar{\omega} = 2\pi f = 500 \times 2\pi = 3141.59 \, \text{rad s}^{-1}$$

(c) $$v = V_p \sin(2\pi ft - kx)$$
$$= 0.1 \sin(2\pi \times 500t - 2\pi x/\lambda)$$
$$= 0.1 \sin(3141.6t - 3612.8)$$

(d) $$v = V_p \sin(2\pi ft - kx)$$
$$= 0.1 \sin\left(\frac{6.28 \times 500 \times 3.25}{10^6} - \frac{6.28 \times 10^3 \times 1.5}{2}\right)$$
$$= 0.1 \sin(0.0102 - 4710)$$
$$= 0.1 \times 0.58 = 0.058 \, \text{V}$$

5.5 Standing waves

A stationary or standing wave occurs when two waves, travelling in opposite directions and having the same frequency, speed and approximately equal amplitudes, are superimposed. This superposition results in points where the displacement is always zero and points where the displacement is always maximum. These points are respectively called **nodes** and **antinodes**. Figure 5.4 shows the formation of a stationary wave which results from the superposition of two progressive waves.

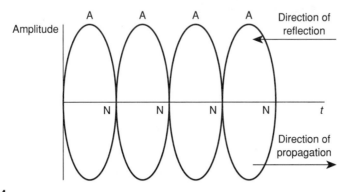

Figure 5.4

5.6 Mathematical treatment of stationary waves

Electromagnetic waves can be reflected and a conducting surface can act as a reflector. The mismatch of cables and aerial feeders produces the same result, but impedance mismatch is generally the cause in this case.

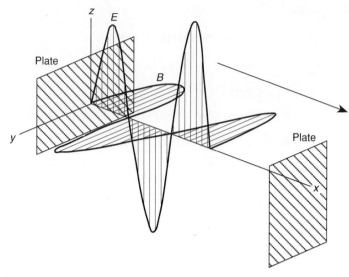

Figure 5.5

If an ideal conductor is placed in the yz-plane as shown in Fig. 5.5 and a linearly polarized electromagnetic wave strikes it as shown, then, as the surface of the conductor is an equipotential surface, the electric field E must be zero everywhere in the yz-plane. Similarly the magnetic field is zero in the yx-plane. Considering the electric field only and taking the incident wave moving from left to right, then

$$e_1 = E \sin(\omega t - kx)$$

Similarly the reflected wave will be given as

$$e_2 = E \sin(\omega t + kx)$$

If these two waves are superimposed the resultant displacement is given as

$$e = e_1 + e_2 = E \sin(2\pi ft - kx) + E \sin(2\pi ft + kx)$$

From the identity

$$\sin(A - B) + \sin(A + B) = 2 \sin A \cos B$$

the expression becomes

$$e = 2E \sin(2\pi ft) \cos(kx) \qquad (5.3)$$

This can be written as

$$e = A \sin 2\pi ft \qquad (5.4)$$

where

$$A = 2E \cos kx \qquad (5.5)$$

Equation (5.4) represents a sinusoidal oscillation of frequency f and whose amplitude A depends on position and is given by equation (5.5). As $k = 2\pi/\lambda$

it can be seen from equation (5.5) that

$$A = 0 \text{ (nodes) when } x = \frac{\lambda}{4}, \frac{3\lambda}{4}, \frac{5\lambda}{4} \ldots \qquad (5.6)$$

$$A = +2E \text{ (antinodes) when } x = 0, \frac{\lambda}{2}, \lambda \ldots \qquad (5.7)$$

If a second conducting plane is inserted in Fig. 5.5 parallel to the first at a distance l from it, then both conducting planes must be nodal planes of E and a standing wave can only exist when the second plane is placed at one of the $E = 0$ positions. Hence for a standing wave to exist l must be a multiple of $\lambda/2$. Hence

$$\lambda_n = \frac{2l}{n} \quad (n = 1, 2, 3 \ldots) \qquad (5.8)$$

5.7 Problems on standing waves

Example 5.6

Electromagnetic waves are set up in a cavity of a multi-cavity klystron tube used in a microwave transmitter. The cavity has two parallel walls separated by 1.2 cm.

(a) Calculate the lowest frequency of the standing waves between the walls.
(b) Determine where the peak magnitude of the electric and magnetic field waves will be in the cavity.

Solution
(a) The lowest frequency corresponds to the longest possible wavelength and this occurs for $n = 1$:

$$\lambda = \frac{2l}{1} = \frac{2 \times 1.2}{1} = 2.4 \text{ cm}$$

hence

$$f = \frac{v}{\lambda} = \frac{3 \times 10^8}{2.4 \times 10^{-2}} = 1.25 \times 10^{10} \text{ Hz}$$

(b) If $n = 1$ there is a single half-wavelength between the walls. The electric field has nodes at the walls and an antinode midway between them, while the magnetic field has antinodes at the walls and a node midway between them.

5.8 Standing waves and the Smith chart

The previous section has highlighted the formation of standing waves along transmission lines. However, of greater importance is the effect that loading the transmission line has on progressive waves.

Loading may be purely resistive, capacitive or inductive or, as is more likely, a combination of these, i.e. impedance loading. Whatever the type of termination, standing waves will be generated along the line if the characteristic impedance Z_0 of the line is not matched to it. In this case the transmission line is fed by a generator while the aerial is taken as the load. When the load and generator impedances are both equal to the characteristic impedance of the line maximum energy is transferred. If the line is terminated in any other resistance standing wave patterns are set up.

A variety of conditions may be obtained by using either an open-circuited or short-circuited transmission line cut to the desired length and standing wave patterns are set up accordingly. The standing wave patterns can be measured by a slotted line which is a device with a probe extending into a transmission line through a slot. The probe can be slid along the slot and by means of a connection to a voltmeter the voltage can be read at any point.

At first glance a resistor placed in parallel or series with the load could produce the necessary matching, but unfortunately a large amount of power would be dissipated. In practice it is customary to use sections of the same transmission line to be matched instead of lumped capacitors and inductors. These sections of transmission line are referred to as matching stubs.

The part played by standing waves is important in calculating input impedance, load impedance and standing wave ratio which is given as

$$S = \frac{V_{max}}{V_{min}} \tag{5.9}$$

All these parameters may be calculated using the Smith chart which generally deals with lossless lines. This is shown in Fig. 5.6. This chart is a plot of normalized impedance, and generally in transmission line problems the practical impedances have to be converted to normalized form before the problem can be tackled.

The Smith chart consists of the following sections:

1. A real axis with values which vary from zero to infinity with unity in the centre.
2. A series of circles with centres on the real axis. These values are normalized.
3. A series of arcs of circles that start from the infinity point on the real axis. Again these are normalized values.

The edge of the chart is marked with scales indicating the coefficient of reflection and distance in wavelengths. Moving around the edge of the chart in a clockwise direction corresponds to moving towards the source or generator end of the line while movement in the opposite direction indicates movement towards the load end of the line. One complete circle represents a half-wavelength.

5.9 Problem solving using the Smith chart

Example 5.7

An aerial having an impedance of $Z_L = 140 + j70$ is connected to a $100\,\Omega$ transmission line. Find the length of the matching stub which will match the aerial to the line and also where it should be placed.

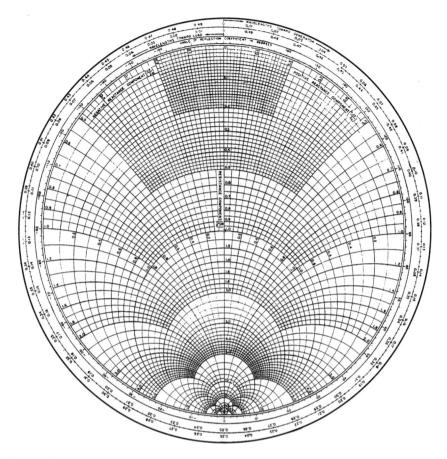

Figure 5.6

Solution

This problem will be tackled in steps:

1. The load impedance is first normalized.
2. This is now plotted on the chart and gives the $S = 2.1$ circle.
3. Draw a line from the normalized load impedance through the centre and project out to the wavelength scale opposite from Z_n. This is found to be 0.439λ (D). Where the projected line crosses the standing wave ratio (SWR) circle indicates the normalized admittance Y_n as at B. The admittance is considered here rather than the impedance, as the final correcting impedance will be placed in parallel with the load at a specific distance from the load. Hence when the correcting circuit is to be placed in parallel with the load, the admittance of the load is considered, whereas when it is placed in series with the load then the impedance is considered. It is, however, the parallel arrangement which is normally considered for a matching stub, hence Y_n is required.
4. Move clockwise around the SWR circle to where it crosses the $R = 1$ circle for the first time at C. This gives the normalized susceptance. This is required

as it is necessary to find the location along the transmission line where the impedance has a resistance component equal to the line impedance and to cancel the reactance at that location. Hence moving around the SWR circle gives the $R = 1$ location so that when 1 is multiplied by Z_0 in the denormalization process the result will be Z_0. The value of the susceptance is $1 + j0.65$.

5. Draw a line from the centre of the chart, through the susceptance point to the wavelength scale. Note that the reading on the wavelengths towards generator scale is 0.149 wavelength approximately (E).

6. The difference in wavelengths between the reading of step 3 and that in step 5 (moving clockwise) is the distance from the load to the point where the matching stub will be placed. In this case it is 0.21 wavelength. This step actually directs a movement from the admittance of the load towards the generator (D to E) to find the point where the impedance on the line is equal to the line impedance itself.

7. The final step is to cancel the susceptance found in step 4. The opposite value of j0.64 is found on the $R = 0$ circle on the bottom of the chart, i.e. $-j0.64$. The reading on the wavelengths towards generator is noted at this point 0.395λ. Finally measure the difference in wavelengths from the point of infinity (0.25λ) to this point (moving from G to F), i.e. $0.395 - 0.25 = 0.145\lambda$. The matching stub is therefore 0.145λ. It is also a shorted stub in order to cut down interference. The graphical solution is shown in Fig. 5.7.

Example 5.8

A radio wave having a frequency of 125 MHz is sent along a loss-free transmission line having a characteristic impedance of 75 Ω. If the transmission line is terminated in a load having an impedance of $(120 + j75)\Omega$, determine, using a Smith chart:

(a) the voltage standing wave ratio;
(b) the distance from the load at which a short-circuit stub should be connected to provide impedance matching;
(c) the length of the stub.

Solution

(a) Since operating frequency is 125 MHz, then

$$\lambda = \frac{v}{f} = \frac{3 \times 10^8}{125 \times 10^6} = 2.4\,\text{cm}$$

Load is $120 + j75$ and this has to be normalized to its normalized value Z_n:

$$Z_n = \frac{120 + j75}{75} = 1.6 + j$$

Plotting this normalized load gives a standing wave ratio of 2.4 and this enables the SWR circle to be drawn. Several points are worth noting about this circle:

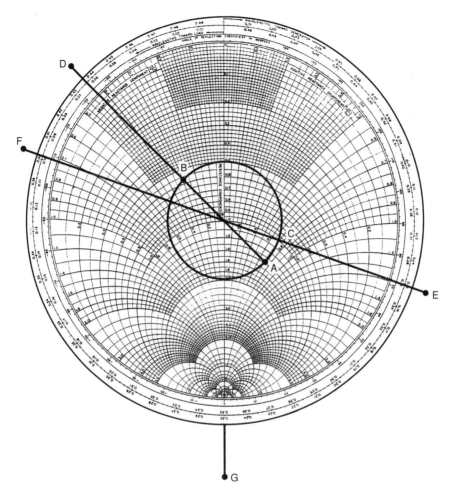

Figure 5.7

(i) The point where the circle crosses the zero reactance line to the right of the centre of the circle indicates the value of the standing wave ratio.

(ii) The value of the normalized load impedance is located on this circle as shown in Fig. 5.8. It also establishes that the standing wave circle is a plot of the impedance along the length of the line.

(iii) The point $X = 0$ on the SWR circle gives the maximum voltage of the standing wave (V_{max}) in wavelengths from the load and the value of the maximum impedance (Z_{max}) at this same point in wavelengths from the load. To the left of the centre of the circle (V_{min}) and (Z_{min}) are also found.

(iv) Normalized susceptance from the chart is $+j0.95$. The difference in readings on the Smith chart gives

$$0.446\lambda - 0.5\lambda = 0.054\lambda$$

hence

$$0.054\lambda + 0.16\lambda = 0.214\lambda$$

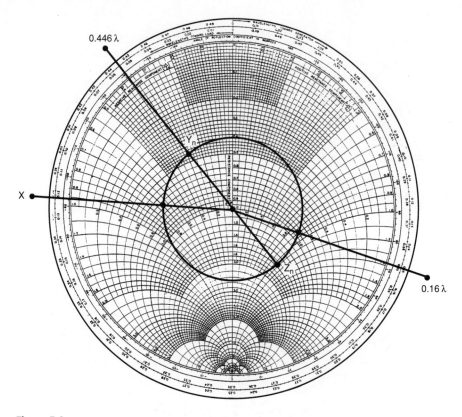

0.446 λ

X

0.16 λ

Figure 5.8

This is the distance from the load terminals to the point where the matching stub will be connected, i.e.

$$0.214 \times 2.4 = 0.5136 \, \text{m}$$

(v) Normalized susceptance $+j0.95$ must now be cancelled. On $R = 0$ circle record $-j0.95$ towards the generator as -0.38λ. Length of stub is

$$0.38\lambda - 0.25\lambda = 0.13\lambda$$

hence

$$0.1 \times 2.4 = 0.312 \, \text{m}$$

Figure 5.8 shows the final solution.

Example 5.9

An aerial is fed by a transmission line as shown in Fig. 5.9. The first voltage minimum is 18 cm from the end of a lossless line terminated in the aerial impedance Z_L. The line is 52 cm long and the SWR $= 2.5$. Adjacent minima are 20 cm apart. Find Z_L and Z_{in}.

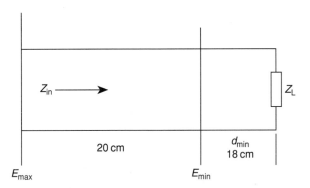

Figure 5.9

Solution
Draw an SWR circle $S = 2.5$. This circle crosses the zero reactance line at a voltage minimum to the left. Also the wavelength on the line is twice the distance between minima, i.e. 40 cm. Since the first voltage minimum occurs 18 cm from the load then this distance in wavelengths is

$$d_{min} = \frac{18}{40}\lambda = 0.45\lambda$$

The next step is to move around the chart 0.45 wavelength towards the load from V_{min}. Draw a line from this point on the wavelength towards the load scale to the point $(1,0)$. The intersection of this line with $S = 2.5$ gives Z_L. From the chart this can be read off as $Z_L - 0.4 + j0.27$. The final step is to find Z_{in} and this may be determined in two ways:

(i) moving $34/40\lambda$ towards the generator from V_{min};
(ii) moving $52/40\lambda$ towards the generator from Z_L.

Move from A to C. This gives $Z_L = 0.89 - j0.89$ on the $s = 2.5$ circle. The complete graphical solution is shown in Fig. 5.10.

Example 5.10

A parallel wire transmission line is terminated in an aerial with an unknown impedance. The first voltage node occurs at 54 cm and the first voltage antinode occurs 204 cm from the load. If $S = 3$ design a short-circuit stub to be placed across the line to match the system.

Solution
Since $V_{max} = 204$ cm and $V_{min} = 54$ cm and there is a quarter wavelength between maximum and minimum values, then

$$\frac{\lambda}{4} = (204 - 54)$$

$$\lambda = 4(204 - 54) = 600 \text{ cm}$$

Using the same procedure as for the previous example gives a short-circuit stub which is 66 cm long and will be placed at 4.2 cm from the load.

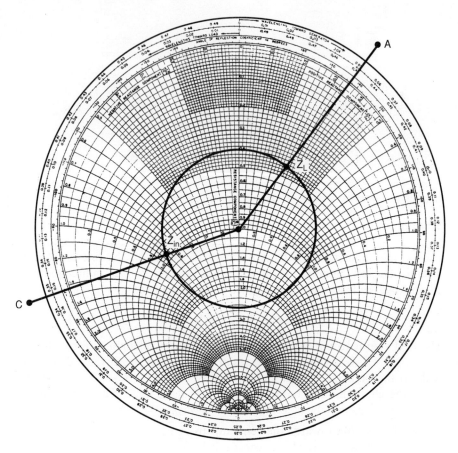

Figure 5.10

Figure 5.11 shows the plotted result. In this case moving from E to D towards the load gives the position of the stub. Moving from G to F gives the length of the stub. Once again the stub is a shorted stub which implies maximum impedance, hence start from Z_{max} and move towards the generator.

Example 5.11

A 50 Ω line has an SWR of 2 when an unknown load impedance is connected to its output terminals. Adjacent voltage minima are found to be 30 cm apart. When the unknown load is removed from the line and is replaced by a short circuit the voltage minima moves by 7.5 cm towards the source. Calculate the value of the unknown load impedance.

Solution
The SWR circle of $S = 2$ is drawn; 30 cm $= \lambda/2$ so that 7.5 cm $= 0.125\lambda$. Moving around the SWR circle for this distance in the clockwise direction gives

$$Z_L = 50(0.8 + j0.58) = 40 + j29 \, \Omega$$

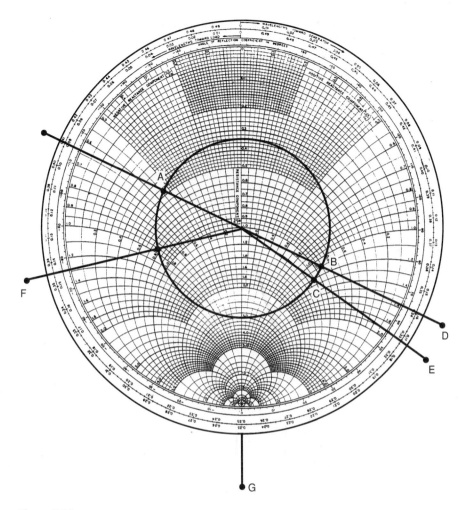

Figure 5.11

5.10 Wave mixing – amplitude modulation

The reception of video, audio or data signals is not possible without the use of a constant-amplitude high-frequency signal. The information is carried by this progressive electromagnetic wave and hence it is called a **carrier** wave. The principle involved is called **modulation** of which there are many types. Two of the most common methods of modulation are considered in this section:

- amplitude modulation
- frequency modulation.

Both these methods can produce complex waveforms, particularly the frequency modulation method. However, it is not the object of this text to treat this topic quantitatively, but it is informative to look at the fundamental role played by progressive waves in this technique and apply the principles to the solution of problems involving telecommunications.

Generally the intelligence and the carrier are combined together in an electronic circuit at the transmitter called a **modulator**. This is shown in the block diagram in Fig. 5.12(b). In practice the information frequency consists of many frequencies, but a single modulating frequency will be considered here. Figure 5.12(a) shows the waveforms.

If the carrier is represented by a progressive wave having no phase shift, then from section 5.2

$$e_c = E_c \sin 2\pi f_c t$$

Similarly the information signal may be represented by

$$e_m = E_m \sin 2\pi f_m t$$

When the two waves are mixed in the modulator the amplitude of the modulated carrier is expressed as

$$(E_c + E_m \sin 2\pi f_m t) \sin 2\pi f_c t$$

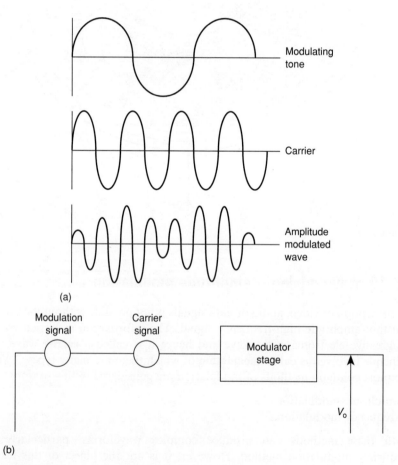

(a)

(b)

Figure 5.12

Expanding this gives

$$E \sin 2\pi f_c t + E_m \sin 2\pi f_m t \sin 2\pi f_c t$$

This is further expanded using trigonometrical identities to give

$$E_c \sin 2\pi f_c t + \tfrac{1}{2} E_m \cos(2\pi f_c - 2\pi f_m)t - \tfrac{1}{2} E_m \cos(2\pi f_c + 2\pi f_m)t$$

The following points should be noted about this expression:

1. $E_c \sin 2\pi f_c t$ is the original carrier frequency.
2. $\tfrac{1}{2} E_m \cos(2\pi f_c - 2\pi f_m)t$ is called the lower sideband and it lags the carrier by 90°.
3. $\tfrac{1}{2} E_m \cos(2\pi f_c + 2\pi f_m)t$ is called the upper sideband and it leads the carrier by 90°.
4. These are two new frequencies which cause the wave envelope in Fig. 5.12.
5. If the carrier is modulated with a range of audio or video signals a range of wavelengths will be present.
6. E_c and E_m are the peak amplitudes of the carrier and modulating signal respectively.
7. The amplitude of the sidebands is half the modulating signal.
8. Adding the two sidebands and carrier gives the total power presented to a common load aerial, i.e.

$$P_T = P_C \left\{ 1 + \frac{m_1^2}{2} + \frac{m_2^2}{2} \right\} \tag{5.10}$$

5.11 Problems on amplitude modulation

Example 5.12

A sinusoidally modulated AM signal is shown in Fig. 5.13. Determine:

(a) the percentage modulation
(b) peak carrier voltage
(c) peak value of the information voltage.

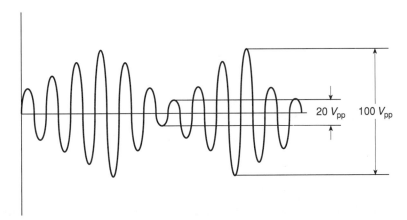

20 V_{pp} 100 V_{pp}

Figure 5.13

Solution

(a) $$m = \frac{100 - 20}{100 + 20} = 0.667 = 66.7 \text{ per cent}$$

(b) The average of the two peak-to-peak measurements is the peak-to-peak amplitude of the unmodulated carrier, i.e. $2V_c$. Hence

$$2V_c = \frac{100 + 20}{2} = 60 V_{pp}$$

(c) $$V_m = mV_c = 0.667 \times 30 = 20 V_p$$

Example 5.13

The instantaneous amplitude of an AM signal is given by

$$v(t) = 120 \left\{ 1 + \sum_{n=1}^{n=4} \frac{1}{2n} \cos(2\pi n 10^4 t) \right\} \cos(2\pi 10^8 t)$$

Calculate:

(a) the peak frequencies and amplitudes of the components of the modulated signal and plot the components on a spectrum graph;
(b) the r.m.s. power that the signal would dissipate in a $70\,\Omega$ load.

Solution

(a) In this type of question the carrier and sidebands should be recognized. The carrier is given as

$$v_c = 120 \cos(2\pi \times 10^8)t$$

First sideband is

$$v_1 = 120 \times \tfrac{1}{2} \cos(2\pi \times 10^4 t) \cos(2\pi \times 10^8 t)$$

Using the trigonometrical identity:

$$\cos A \cos B = \tfrac{1}{2} \{ \cos(A + B) + \cos(A - B) \}$$

$$v_1 = 120 \times \tfrac{1}{2} \times \tfrac{1}{2} \{ \cos(2\pi \times 10^4 + 2\pi \times 10^8)t \}$$
$$+ \{ \cos(2\pi \times 10^4 - 2\pi \times 10^8)t \}$$
$$= 30 \text{ V}$$

Similarly, the other sidebands are given as

Second sideband $v_2 = 120 \times \tfrac{1}{4} \times \tfrac{1}{2} = 15 \text{ V}$

Third sideband $v_3 = 120 \times \tfrac{1}{6} \times \tfrac{1}{2} = 10 \text{ V}$

Fourth sideband $v_4 = 120 \times \tfrac{1}{8} \times \tfrac{1}{2} = 7.5 \text{ V}$

Hence the spectrum diagram showing all the sideband frequencies is shown in Fig. 5.14.

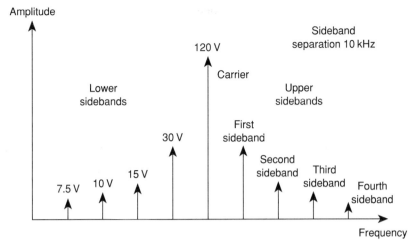

Figure 5.14

(b) The r.m.s. power is generally considered for transmission problems and it simply means 'the square root value of the mean of the squares' of all the voltage levels for the upper and lower sidebands. Hence

$$V_{rms} = \frac{\sqrt{120^2 + 2 \times (15)^2 + 2 \times (10)^2 + 2 \times (7.5)^2}}{2}$$

$$= 87.1\,\text{V}$$

So

$$P = \frac{87.1^2}{70} = 108.4\,\text{W}$$

Example 5.14

A carrier wave of frequency 1 MHz and amplitude $10V_{pp}$ is amplitude modulated by a sinusoidal modulating signal. If the lower sideband is 980 kHz and its voltage is 15 dB below the carrier amplitude, calculate:

(a) the peak-to-peak amplitude of the modulating signal;
(b) the frequency of the modulating signal.

Solution

(a) The concepts of **decibels** is introduced here. The decibel is a logarithmic intensity level which follows the logarithmic responses of the ear. It is frequently applied to wave amplitudes and is applied in two forms depending on whether voltage or power levels are being considered:

$$dB = 20\log_{10}\frac{V^1}{V^2} \quad \text{(voltage form)}$$

$$dB = 10\log_{10}\frac{P_1}{P_2} \quad \text{(power form)}$$

As voltages are being considered here,

$$15 = 20\log\left(\frac{\text{carrier voltage}}{\text{sideband voltage}}\right) = 20\log\frac{V_c}{V_s}$$

$$0.75 = \log_{10}\left(\frac{\text{carrier voltage}}{\text{sideband voltage}}\right)$$

Note that the smaller value, namely the sideband voltage, is placed on the denominator to give a positive value. Taking antilogs of both sides gives

$$100^{0.75} = \frac{V_c}{V_s}$$

$$5.65 = \frac{V_c}{V_s}$$

So

$$V_s = \frac{V_c}{5.65} = \frac{10}{5.65} = 1.77\,\text{V}$$

Since the amplitude of the sideband is equal to one-half of the voltage of the modulating signal, the modulating signal amplitude is $3.54V_{pp}$.

(b) The frequency of the lower sideband is

$$10^6 - 98 \times 10^4 = 20\,\text{kHz}$$

Example 5.15

A communications system uses an unmodulated carrier with a power of 2 kW. This carrier is modulated by two test tones producing depths of modulation corresponding to 25 and 36 per cent respectively. Determine:

(a) the total sideband power
(b) the upper sideband power.

Solution
Since two tones are used:

$$P_T = P_C\left(1 + \frac{(m_1)^2}{2} + \frac{(m_2)^2}{2}\right)$$

So

$$P_T = 2000 + 62.5 + 129.6 = 2192.1\,\text{W}$$

Total sideband power is 192.1 W, hence the upper sideband power is half of this, i.e. 96 W.

Example 5.16

A carrier and modulating signal are applied in series to a modulator (Fig. 5.15) which has a characteristic given by

$$V_{out} = 0.6V_{in} + 0.2V_{in}^2$$

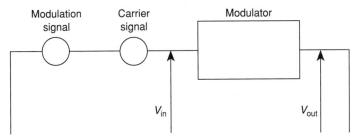

Figure 5.15

If the two signals are fed in series determine all the frequency components and show them on a spectrum diagram.

Solution

Since the two signals are fed in series then the complex signal fed to the modulator will be

$$V_{in} = 14 \sin 2\pi \times 10^5 t + 2 \sin 4\pi \times 10^3 t$$

$$V_{in}^2 = 196 \sin^2 2\pi \times 10^5 t + 4 \sin^2 4\pi \times 10^3 t$$

$$= 56 \sin 2\pi \times 10^5 t \sin 4\pi \times 10^3$$

Since $V_{out} = 0.6 V_{in} + 0.2 V_{in}^2$ then substituting the previous expressions into this expression gives

$$V_{out} = 8.4 \sin 2\pi \times 10^5 t + 1.2 \sin 4\pi \times 10^3 t$$

$$+ 39.2 \sin^2 2\pi \times 10^5 t + 0.8 \sin^2 4\pi \times 10^3 t$$

$$+ 11.2 \sin 2\pi \times 10^5 t \sin 4\pi \times 10^3$$

Using the trigonometrical identity

$$\sin A \sin B = 1\{\cos(A - B) - \cos(A + B)\}$$

will give

$$V_{out} = 8.4 \sin 2\pi \times 10^5 t + 1.2 \sin 4\pi \times 10^3 t$$

$$+ 19.6(1 - \cos 4\pi \times 10^5 t)$$

$$+ 0.4(1 - \cos 8\pi \times 10^3 t)$$

$$+ 5.6\{\cos(2\pi \times 10^5 t - 4\pi \times 10^3 t) - \cos(2\pi \times 10^5 t + 4\pi \times 10^3 t)\}$$

Since $f_c = 100\,\text{kHz}$ and $f_m = 2\,\text{kHz}$ then the spectrum diagram is given as Fig. 5.16.

5.12 Wave mixing – frequency modulation

This technique is more complex and in this case rather than the amplitude of the wave being changed the frequency of the wave is changed. This is shown in

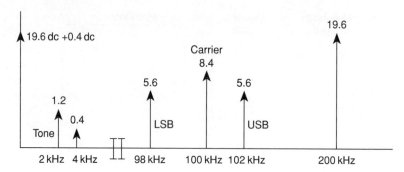

Figure 5.16

Fig. 5.17. It is noticeable, in this case, that as the amplitude of the modulating frequency is increased the frequency swings or changes. This is known as the **frequency deviation**.

Frequency modulation can also be applied to square wave. modulating signals, in which case the amplitude of the square wave is constant and therefore the frequency change is constant. Square waves will be looked at later.

In practical transmission systems the maximum frequency deviation which is permitted to occur for a particular system is called the **rated system deviation**. One other parameter used in frequency modulation is the **deviation ratio** and this is given as follows:

$$\delta = \frac{\text{Rated system deviation}}{\text{Maximum modulating frequency}} = \frac{f_d}{f_m} \qquad (5.11)$$

As will be seen later, this parameter enables the sidebands to be determined. This is done by the use of Bessel functions which are given in either tabular or graphical form. Generally, however, it is more convenient to work from tables such as those given in Appendix A.

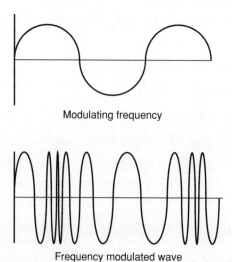

Modulating frequency

Frequency modulated wave

Figure 5.17

A final problem is how to accommodate all the sidebands in a particular part of the spectrum which is technically called the **bandwidth** (BW). This condition has also to be satisfied for amplitude modulation, but with frequency modulation a greater bandwidth is required because of the larger number of sidebands. In order to calculate this bandwidth, a rule known as **Carson's rule** is generally used to give an approximation. It is given in terms of the rated system deviation and the maximum modulating frequency:

$$BW = 2(f_d + f_m) \tag{5.12}$$

where BW is the bandwidth, f_d the rated system deviation and f_m the maximum modulating frequency.

5.13 Problems on frequency modulation

Example 5.17

The rated system deviation of an FM system is 50 kHz for a modulation amplitude of 10 V. Use the information given on Bessel tables to determine the frequency spectrum when the carrier is modulated in turn by audio signals of

(a) 3 V amplitude at 165 kHz
(b) 4 V amplitude at 8 kHz.

The carrier frequency is 80 MHz and the carrier amplitude is 100 V.

Solution
In this problem the rated system deviation occurs for a modulation level of 10 V, hence the deviation ratio in both cases is given as

$$\beta_1 = \frac{k_1 f_d}{f_{m_1}} = \frac{\frac{3}{10} \times 50 \times 10^3}{15 \times 10^8} = \frac{150}{150} = 1$$

$$\beta_2 = \frac{k_2 f_d}{f_{m_2}} = \frac{\frac{4}{10} \times 50 \times 10^3}{8 \times 10^3} = \frac{200}{80} = 2.5$$

From the Bessel tables the two sets of values can be obtained:

$$J_0 = 0.7652 \quad J_1 = 0.4401 \qquad J_0 = 0.05 \quad J_1 = 0.5$$
$$J_2 = 0.1149 \quad J_3 = 0.0196 \qquad J_2 = 0.45 \quad J_3 = 0.2$$
$$J_4 = 0.08 \quad J_5 = 0.02$$

The frequency spectrum now becomes that shown in Fig. 5.18.

Example 5.18

An FM system has a rated system deviation of 75 kHz and this occurs when a modulating frequency of 15 kHz and 6 V amplitude is applied to the carrier of

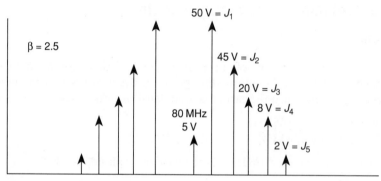

Figure 5.18

92 MHz. Determine the frequency/voltage spectrum if the peak carrier voltage
is 10 V and only sidebands greater than or equal to 10 per cent of the unmodulated carrier voltage are transmitted.

Solution
The deviation is

$$\delta = \frac{75}{15} = 5$$

Hence from the Bessel tables the following values are obtained:

$$J_0 = 0.1776 = -1.776\,\text{V}$$

$$J_1 = 0.3276 = -3.276\,\text{V}$$

$$J_2 = 0.0466 = 0.466\,\text{V}$$

$$J_3 = 0.3648 = 3.648\,\text{V}$$

$$J_4 = 0.3912 = -3.912\,\text{V}$$

$$J_5 = 0.2611 = -2.611\,\text{V}$$

$$J_6 = 0.1310 = 1.310 \, \text{V}$$

$$J_7 = 0.534 \ = 5.34 \, \text{V}$$

$$J_8 = 0.0184 = 0.184 \, \text{V}$$

These values can now be plotted on a spectrum diagram.

Example 5.19

An FM transmitter has a carrier frequency of 3.5 MHz produced at the oscillator. Three tripler sections are used to multiply the frequency. The amplitude of the oscillator frequency is 8 V and the modulator has a sensitivity of 1.5 kHz/V. If the modulating signal is given by

$$V_m = 1.42 \sin(120 \times 10^3)t$$

Calculate:

(a) the modulating index;
(b) the amplitude of all sidebands up to an amplitude 10 per cent of the unmodulated carrier voltage;
(c) the required bandwidth.

Solution

(a) The centre frequency after frequency multiplication is

$$3.5 \times 3 \times 3 \times 3 = 94.5 \, \text{MHz}$$

The frequency deviation is determined by using the sensitivity of the modulator and also the multiplication effect of the three triplers. Hence taking the amplitude of the modulating signal as 1.42 V:

$$1.5 \times 27 \times 1.42 \times 10^3 = 57.51 \, \text{MHz}$$

Hence the modulation index is

$$\delta = \frac{1.5 \times 27 \times 1.42 \times 10^3}{120 \times 10^3 / 2\pi} = 3$$

(b) The sidebands can now be determined from the Bessel tables as before. This gives a frequency spectrum diagram (Fig. 5.19).
(c) By Carson's rule:

$$BW = 2(f_d + f_m) = 2(57.51 + 19.108) = 153.2 \, \text{kHz}$$

5.14 Harmonics

When a string is plucked in a stringed instrument a fundamental tone is produced together with multiples of this fundamental. These are called the **harmonics** of the fundamental and are responsible for the quality of the sound produced. It has already been shown that the bandwidth of a system

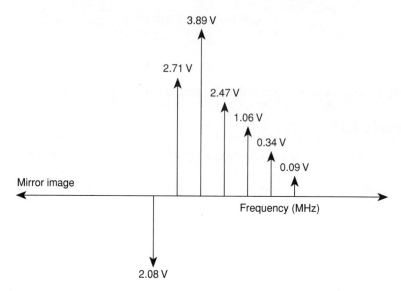

Figure 5.19

should contain all useful wavelengths, e.g. sidebands. A similar requirement is needed when harmonics are produced and several applications are stated below.

- Telephone cables are limited in their frequency response (known as the **transient response**). A cable should have as large a capacity as possible in order to accommodate a large number of channels. If this bandwidth is not large then the amount of information is restricted.
- Broadcasting stations are licensed to operate in assigned frequency channels in order to avoid interference.
- Square wave and video amplifiers should have a large bandwidth as a square wave consists of many harmonics which must be passed by the amplifier in order to retain the quality of the information.
- Harmonics have a detrimental effect on power audio amplifiers as they cause harmonic distortion of the output waveforms. Generally, second harmonic distortion is the worse type.
- Harmonics are produced in nonlinear devices in electronic circuits and generally have to be filtered.

The simple progressive wave discussed in section 5.1 really indicates a fundamental wave or **first harmonic**. Harmonics of this wave may be given as

$$A_2 \sin 2\pi 2ft \quad \text{(second harmonic)} \tag{a}$$

$$A_3 \sin 2\pi 3ft \quad \text{(third harmonic)} \tag{b}$$

$$A_4 \sin 2\pi 4ft \quad \text{(fourth harmonic)} \tag{c}$$

A wave with a fundamental and second harmonic is shown in Fig. 5.20.

Most information is currently transmitted digitally and for this reason a knowledge of the frequency components in transmitted pulses is desirable for

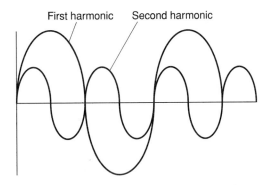

First harmonic   Second harmonic

Figure 5.20

the sake of signal processing and circuit response. Any wave function can be resolved into a number of sinusoids and the method of doing this is by Fourier analysis. In practice an instrument called a spectrum analyser performs this function.

5.15 Fourier analysis

A full mathematical analysis of the Fourier transform confirms the following results:

$$a_o = \frac{1}{T} \int_0^T f(t)\, dt \qquad\qquad (5.13)$$

$$a_n = \frac{2}{T} \int_0^T f(t) \cos 2\pi nft\, dt \qquad\qquad (5.14)$$

$$b_n = \frac{2}{T} \int_0^T f(t) \sin 2\pi nft\, dt \qquad\qquad (5.15)$$

Equation (5.13) is termed the d.c. component while equations (5.14) and (5.15) evaluate the harmonic magnitudes. Many waveforms may be analysed by this method and these will be shown later.

5.16 Analysis of a square wave

The output of a sinusoidal oscillator is almost pure if the circuit is designed properly with the correct feedback, but often harmonics are present and hence a Fourier series can be derived in the form already shown in expressions (a), (b) and (c) on p. 196.

In digital electronic systems the pulse or square wave is more important, and for this reason an analysis of this particular waveform is given here. Other waveforms are used and can be tackled in the same way by using expressions (5.13)–(5.15).

Example 5.20

The function in Fig. 5.21 is defined over one period as follows:

$$v(t) = \left\{\begin{matrix} A \\ 0 \end{matrix}\right\}\left\{\begin{matrix} 0 \leq t \leq T/2 \\ T/2 \leq t \leq T \end{matrix}\right\}$$

$$a_o = \frac{1}{T}\int_0^{T/2} A \, dt = \left.\frac{A_t}{T}\right|_0^{T/2} = \frac{A}{T}(T/2)$$

$$= \frac{A}{2} \quad \text{(the average d.c. value)}$$

$$a_n = \frac{2}{T}\int_0^{T/2} A \cos 2\pi nft \, dt = \frac{2A}{2\pi nf\,T} \sin 2\pi nf \times T/2$$

Since $f = 1/T$:

$$a_n = \frac{A}{\pi n} \sin \pi n = 0 \quad \text{(for all } n\text{)}$$

$$b_n = \frac{T}{2}\int_0^{T/2} A \sin 2\pi nft \, dt$$

$$= -\frac{A}{\pi nf\,T}(\cos 2\pi nf \times T/2 - \cos 0)$$

$$= \frac{A}{\pi n}(1 - \cos \pi n)$$

$$= \frac{A}{\pi n}(1 + 1) \quad \text{(for } n \text{ odd)}$$

and

$$= \frac{A}{\pi n}(1 - 1) \quad \text{(for } n \text{ even)}$$

Hence

$$b_r = \frac{2A}{n\pi} \quad \text{(} n \text{ odd only)}$$

So the complete expression may be written as

$$v(t) = \frac{A}{2} + \sum_{n\,\text{odd}}^{\infty} \frac{(2A)}{\pi n} \sin 2\pi nft$$

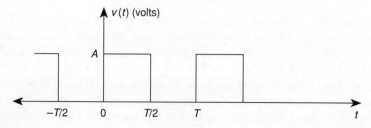

Figure 5.21

Familiarity with odd and even functions would have shown that only sine functions would be present. It can be concluded that a square wave can be produced with a series of sine wave generators.

5.17 Signal processing

There are many applications where waveform conditioning is required. Often the use of filtering is used to process the signal, and in order to design the filter a knowledge of the frequency components in the waveform is essential.

Sinusoidal waves may have several harmonic elements and in situations such as power amplifiers, where second harmonic distortion is prevalent, methods are used to reduce this effect, and hence produce a distortionless sound.

This section will concentrate on the square wave which in many cases requires a wideband amplifier in order to accept all the components and hence produce a well-defined square wave.

Example 5.21

It is informative to consider the effect an ideal filter has on a square wave having certain odd harmonics (Fig. 5.22).

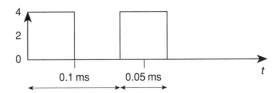

Figure 5.22

Solution
This waveform is passed through an ideal filter with a gain of unity and having the response shown in Fig. 5.23.

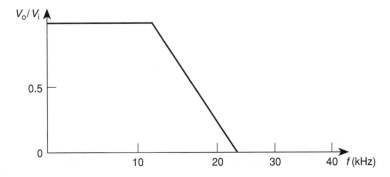

Figure 5.23

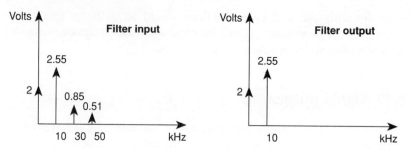

Figure 5.24

From the result of Example 5.20 the frequency components of a 10 kHz/4 V square wave are given as

$$v(t) = 2 + \frac{8}{\pi} \sin 2\pi \times (10)t + \frac{8}{3\pi} \sin 2\pi(30)t$$

$$+ \frac{8}{5\pi} \sin 2\pi(50)t + \ldots \text{volts}$$

The input signal spectrum and the filter output spectrum are shown in Fig. 5.24. Remember this is for an ideal filter which does not allow for any roll-off. The response of this filter is to pass a zero frequency or d.c. level of 2 V and also a 10 kHz signal at $2.55V_p$. All the other components are suppressed due to the gain of the filter falling off. Shown graphically this would indicate a 10 kHz sinusoidal signal fluctuating on a 2 V d.c. level.

Example 5.22

A 50 Hz sinusoidal supply uses a half-wave rectifier. Determine the d.c. voltage and the peak voltage of the output 50 and 100 Hz components. Assume the peak amplitude is unchanged at 10 V and that no filter capacitor is used.

Solution
The Fourier series for this waveform is obtained from the table and is given as

$$v(t) = \frac{A}{\pi} + \frac{A}{2} \sin 2\pi(f)t - \frac{2A}{3\pi} \cos 2\pi(2f)t \ldots \text{volts}$$

$$v(t) = \frac{10}{\pi} + \frac{10}{2} \sin 2\pi(50)t - \frac{20}{3\pi} \cos 2\pi(100)t \ldots \text{volts}$$

Evaluating these results will give the d.c. value of this supply as well as the fundamental and second harmonic values. This is shown on the spectrum diagram in Fig. 5.25.

Example 5.23

An amplifier is saturated resulting in the output shown in Fig. 5.26.

(a) Sketch the amplifier output spectrum diagram.

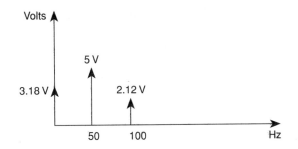

Figure 5.25

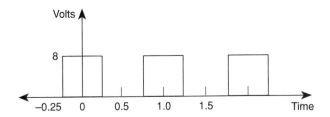

Figure 5.26

(b) If a filter with the response shown in Fig. 5.27 is connected to the amplifier, sketch the spectrum diagram at the output.

(c) Determine the total harmonic distortion at the output of the amplifier.

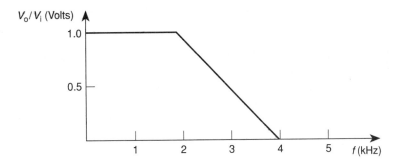

Figure 5.27

Solution

(a) From the Fourier table in Appendix B the solution is

$$v(t) = \frac{A}{2} + \sum_{n=1}^{\infty} \frac{A\sin(n\pi/2)}{n\pi/2} \cos(n2\pi ft)$$

$$= \frac{A}{2} + \frac{2A}{\pi} \cos 2\pi(f)t + \frac{2A}{3\pi} \cos 2\pi(3f)t$$

$$+ \frac{2A}{3\pi} \cos 2\pi(5f)t \ldots \text{volts}$$

This gives a d.c. value of 5 V, a fundamental of 1 kHz, a third harmonic of 3 kHz and a fifth harmonic of 5 kHz. The respective voltage values are $6.4V_p$, $2.1V_p$ and $1.3V_p$.

(b) When this is passed through the filter only the 3 kH waveform and the d.c. component are present together with the fundamental. These values are the same as before except for the third harmonic which will now have a value of $1.05V_p$ due to the response of the filter.

(c) The total harmonic distortion (THD) is given by

$$\text{THD} = \sqrt{d_3^2 + d_5^2 \ldots}$$

Hence percentage total harmonic distortion is

$$\%\text{THD} = \frac{\sqrt{(2.1)^2 + (1.3)^2}}{6.4} = 38.5\%$$

5.18 Further problems

1. A voltage wave travels along a transmission line in the positive x-direction. If $\omega = 6.66 \times 10^3$ rad/s, $\lambda = 0.3$ m and the phase angle is given as $-2.4°$ determine:
   (a) the speed of the wave;
   (b) the value of the x-displacement;
   (c) the corresponding value of t for the x-displacement.
   Answer: 315 m/s; 13.23 m; 6.34 μs

2. A sinusoidal electromagnetic force having a frequency of 60 Hz and a peak amplitude of 300 V is applied to a circuit. Calculate:
   (a) the EMF after 2 ms;
   (b) the time taken for the EMF to reach 250 V in the positive direction;
   (c) the time taken for the EMF to reach 150 V in the negative direction.
   Answer: 205 V; 2.61 ms; 9.72 ms

3. Two alternating voltages are connected in series in a circuit. They are represented as follows:

$$e_1 = 10\sin 628t$$

$$e_2 = 5\sin(628t + \pi/4)$$

Determine the resultant voltage and its phase displacement relative to e_1. Note this problem can be solved by graphical addition or, as is more general, by phasor diagram as the frequency of both waveforms is the same.
Answer: 14 V, 15°

4. The equation:

$$v = V_p \sin(\omega t - kx)$$

where $k = 2\pi/\lambda$ represents a wave travelling along the x-direction. If $\omega = 8 \times 10^3$ rad s^{-1}, $k = 25$ and $x = 3.5$ mm, calculate:
(a) the speed of the wave;
(b) the time displacement of a particle at x;
(c) the phase angle.
Answer: 319 m s^{-1}; 11 μs; 5°

5. Two sinusoidal voltages are transmitted along a transmission line. They are represented by the following equations:

$$v_1 = 150 \sin(\theta - 40°)$$

$$v_2 = 200 \sin(\theta + 70°)$$

Determine:
(a) the resultant voltage travelling along the line by algebraically adding the waves;
(b) the r.m.s. and average values of the resultant wave.
Answer: $212 \sin(\theta + 20°)$; 149.9 V; 134.96 V

6. A load of $90 - j120\,\Omega$ terminates a $50\,\Omega$ lossless transmission line. Find by use of a Smith chart:
(a) the voltage standing wave ratio (VSWR);
(b) the distance from the load to the first voltage minimum if the operating frequency is 750 MHz.
Answer: 5.4; 8.28 cm

7. A transmission line having $Z_0 = 80\,\Omega$ feeds into an aerial with $Z_L = 36 - j15$. If the operating frequency is 459 MHz, determine with the use of a Smith chart:
(a) the standing wave ratio;
(b) the distance from the load that a matching stub must be placed;
(c) the length of the shorted stub.
Answer: 2.3; 8.2 cm; 23.96 cm

8. A lossless transmission line of $Z_0 = 100\,\Omega$ is terminated by an unknown impedance. The termination is found to be at a maximum of the voltage standing wave. If the VSWR is 5, determine the value of the terminating impedance.
Answer: 500 Ω

9. Measurements are carried out on a line and the following is noted: VSWR = 3.5; minima are 15 cm apart and $V_{min} = 1.2$ cm. A short-circuit shunt stub has to be used for the matching of the line. Determine:
(a) the length of the stub;
(b) the position of the stub.
Answer: Either 3.12 cm long placed 13.83 cm from the load or 11.88 cm long placed 2.49 cm from the load.

10. A radio wave having a frequency of 95 MHz is propagated in a loss-free transmission line. The transmission line is terminated in a load having an

impedance of $225 - j75$. Given that the characteristic impedance of the transmission line is $75\,\Omega$, determine from the Smith chart:

(a) the VSWR;

(b) the distance from the load at which a short-circuit stub would be connected to provide impedance matching;

(c) the reactance and length of the stub.

Answer: 3.5; 0.1911λ; $0.322\,\mathrm{m}$; $+j0.74\,\Omega$

11. If a transmitter radiates $4\,\mathrm{kW}$ with an unmodulated carrier wave and $5.2\,\mathrm{kW}$ when the carrier wave undergoes a variation in amplitude, calculate the percentage of modulation.

Answer: 77.7 per cent

12. An AM modulator has the following characteristic:

$$V_{\mathrm{out}} = 0.8V_{\mathrm{in}} + 0.2V_{\mathrm{in}}^2$$

It is driven by the following modulated signal:

$$V_{\mathrm{in}} = 12\sin 6.28 \times 10^6 t + 2\sin 31.42 \times 10^3 t$$

(a) Draw a frequency spectrum diagram of the modulator output.

(b) Calculate the modulation depth assuming that the unwanted frequency components can be rejected.

Answer: Select upper and lower sidebands at $1.005\,\mathrm{MHz}$ and $0.995\,\mathrm{MHz}$ for spectrum diagram; 100 per cent

13. The instantaneous amplitude of an AM signal is given by the expression:

$$v(t) = 100\left\{1 + \sum_{n-1}^{3} \frac{1}{2n}\cos(2\pi n 10^3 t)\right\}\cos(2\pi 10^7 t) \text{ volts}$$

Determine the peak amplitudes and frequencies of the various components of the modulated signal and plot the voltage amplitude/frequency spectrum.

Answer: 25, 12.5, 8.3 V. All sidebands separated by $1\,\mathrm{kHz}$.

14. An FM station is allocated a channel 91.4–$91.6\,\mathrm{MHz}$. The maximum permitted frequency deviation is $75\,\mathrm{MHz}$. Sidebands of amplitude less than 1 per cent of unmodulated carrier amplitude are not transmitted.

Using Bessel functions find the maximum permitted modulation index of an $8\,\mathrm{kHz}$ sinusoidal test signal, and the frequency deviation caused by this signal.

Answer: 8; sideband levels are: 1.173, 0.929, 1.69, 1.6, 1.1, 0.63, 0.03, 1, -0.565, -1.455, $-0.525\,\mathrm{V}$

15. A $5V_{\mathrm{p}}$ $100\,\mathrm{MHz}$ carrier is to be frequency modulated by a $10\,\mathrm{kHz}$ sinusoidal signal such that the peak frequency deviation is $30\,\mathrm{kHz}$. If only side frequencies with amplitudes greater than 10 per cent of the unmodulated carrier are to be transmitted, sketch the frequency spectrum of the transmitted wave and determine the total bandwidth required.

Answer: $\delta = 7.5$; $152\,\mathrm{kHz}$

16. When no signal is applied to a modulator in a transmitter, the carrier output has a frequency of 100 MHz and an amplitude of $10V_p$. An input signal causes a frequency deviation of 20 kHz/V. If the modulating signal applied is

$$V_m = 1.75\sin(62.8 \times 10^3)t$$

draw the spectrum diagram.
Answer: Modulation index $= 3.5$; amplitudes of sidebands are: 1.4, 4.6, 3.8, 2.1, 0.8 V separated by 1 kHz

17. A 0.05 ms, $3V_p$ rectangular pulse train centred at the origin with a period of 1 ms is applied to a filter. If the frequency response of the filter is given as shown in Fig. 5.28:
    (a) determine the d.c. voltage and peak amplitudes of the first five sinusoidal components at the input;
    (b) sketch the waveform which would appear on an oscilloscope connected to the output of the filter.
    Answer: (a) 0.75, 0.35, 0.955
            (b) 1 kHz signal fluctuating on a 1.5 V d.c. level

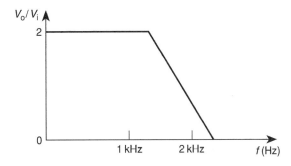

Figure 5.28

18. A full-wave unregulated supply gives a peak amplitude of 6 V. Determine the peak voltages of the d.c. and first three sinusoidal components of this output and sketch them on a spectrum diagram.
    Answer: 3.82; 2.55; 0.04; 0.009

19. A linear amplifier has a 2.5 kHz tone applied to its input and the amplifier itself generates a second and third harmonic. If there is no d.c. component present:
    (a) determine the percentage of second and third harmonic distortion if the voltage levels are respectively:

$$V_1 = 5\,\text{V}; V_2 = 2\,\text{V}; V_3 = 1\,\text{V}$$

    (b) sketch the complex waveform showing the three components of the output waveform.
    Answer: 40 per cent; 20 per cent

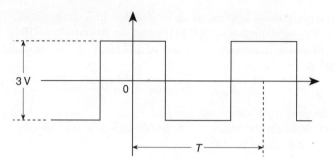

Figure 5.29

20. A square wave symmetrical about the origin has, as shown in Fig. 5.29, a period of 0.2 ms and an amplitude of 3 V. Determine the first four components and draw a sinusoidal diagram showing these components of the square wave.

Answer: 5 kHz, 0.095 V; 15 kHz, −0.32 V; 25 kHz, 0.19 V; 35 kHz, −0.005 V

Appendix A Table of Bessel functions

Bessel function values for modulation indices up to 15

n	$J_n(1)$	$J_n(2)$	$J_n(3)$	$J_n(4)$	$J_n(5)$	$J_n(6)$	$J_n(7)$	$J_n(8)$	$J_n(9)$	$J_n(10)$	$J_n(11)$	$J_n(12)$	$J_n(13)$	$J_n(14)$	$J_n(15)$
0	0.7652	0.2239	−0.2601	−0.3971	−0.1776	0.1506	0.3001	0.1717	−0.0903	−0.2459	−0.1712	0.0477	0.2069	0.1711	−0.0142
1	0.4401	0.5767	0.3391	−0.0660	−0.3276	−0.2767	−0.0047	0.2346	0.2453	0.0435	−0.1768	−0.2234	−0.0703	0.1334	0.2051
2	0.1149	0.3528	0.4861	0.3641	0.0466	−0.2429	−0.3014	−0.1130	0.1448	0.2546	0.1390	−0.0849	−0.2177	−0.1520	0.0416
3	0.0196	0.1289	0.3091	0.4302	0.3648	0.1148	−0.1676	−0.2911	−0.1809	0.0584	0.2273	0.1951	0.0033	−0.1768	−0.1940
4	−	0.0340	0.1320	0.2811	0.3912	0.3576	0.1578	−0.1054	−0.2655	−0.2196	−0.0150	0.1825	0.2193	0.0762	−0.1192
5	−	−	0.0430	0.1321	0.2611	0.3621	0.3479	0.1858	−0.0550	−0.2341	−0.2383	−0.0735	0.1316	0.2204	0.1305
6	−	−	0.0114	0.0491	0.1310	0.2458	0.3392	0.3376	0.2043	−0.0145	−0.2016	−0.2437	−0.1180	0.0812	0.2061
7	−	−	−	0.0152	0.0534	0.1296	0.2336	0.3206	0.3275	0.2167	0.0184	−0.1703	−0.2406	−0.1508	0.0345
8	−	−	−	−	0.0184	0.0565	0.1280	0.2235	0.3051	0.3179	0.2250	0.0451	−0.1410	−0.2320	−0.1740
9	−	−	−	−	−	0.0212	0.0589	0.1263	0.2149	0.2919	0.3089	0.2304	0.0670	−0.1143	−0.2200
10	−	−	−	−	−	−	0.0235	0.0608	0.1247	0.2075	0.2804	0.3005	0.2338	0.0850	−0.0901
11	−	−	−	−	−	−	−	0.0256	0.0622	0.1231	0.2010	0.2704	0.2927	0.2357	0.0999
12	−	−	−	−	−	−	−	−	0.0274	0.0634	0.1216	0.1953	0.2615	0.2855	0.2367
13	−	−	−	−	−	−	−	−	0.0108	0.0290	0.0643	0.1201	0.1901	0.2536	0.2787
14	−	−	−	−	−	−	−	−	−	0.0119	0.0304	0.0650	0.1188	0.1855	0.2464
15	−	−	−	−	−	−	−	−	−	−	0.0130	0.0316	0.0656	0.1174	0.1813
16	−	−	−	−	−	−	−	−	−	−	−	0.0140	0.0327	0.0661	0.1162
17	−	−	−	−	−	−	−	−	−	−	−	−	0.0149	0.0337	0.0665
18	−	−	−	−	−	−	−	−	−	−	−	−	−	0.0158	0.0346
19	−	−	−	−	−	−	−	−	−	−	−	−	−	−	0.0166

Note: Only those values greater than 0.0100 are given.

Bessel function values for modulation indices less than unity

n	$J_n(0.1)$	$J_n(0.2)$	$J_n(0.3)$	$J_n(0.4)$	$J_n(0.5)$	$J_n(0.6)$	$J_n(0.7)$	$J_n(0.8)$	$J_n(0.9)$	$J_n(1.0)$
0	0.9975	0.9900	0.9776	0.9604	0.9385	0.9120	0.8812	0.8463	0.8075	0.7652
1	0.0499	0.0995	0.1483	0.1960	0.2423	0.2867	0.3290	0.3686	0.4059	0.4401
2	–	–	0.0112	0.0197	0.0306	0.0437	0.0588	0.0758	0.0946	0.1149
3	–	–	–	–	–	–	–	0.0102	0.0144	0.0196

Note: Only those values greater than 0.0100 are given.

Appendix B Fourier transforms

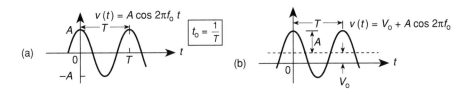

(a)

(b)

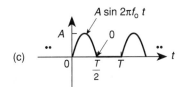

(c)

$$v(t) = \frac{A}{\pi} + \frac{A}{2} \sin 2\pi f_0 t - \frac{2A}{3\pi} \cos 2\pi(2f_0)t - \frac{2A}{15\pi} \cos 2\pi(4f_0)t + \dots$$

$$= \frac{A}{\pi} + \frac{A}{2} \sin 2\pi f_0 t + \sum_{n=2}^{\infty} \frac{A[1 + (-1)^n]}{\pi(1 - n^2)} \cos 2\pi(nf_0)t$$

(d)

* (the rectifier input signal
will have a period of 2 T)

$$v(t) = \frac{2A}{\pi} + \frac{4A}{3\pi} \cos 2\pi f_0 t - \frac{4A}{15\pi} \cos 2\pi(2f_0)t + \dots$$

$$= \frac{2A}{\pi} + \sum_{n=1}^{\infty} \frac{4A(-1)^n}{\pi[1 - (2n)^2]} \cos 2\pi(nf_0)t$$

(e)

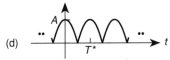

$$v(t) = \frac{2A}{\pi} \sin 2\pi f_0 t + \frac{2A}{3\pi} \sin 2\pi(3f_0)t + \dots$$

$$= \sum_{n,\,\text{odd only}}^{\infty} \frac{2A}{n\pi} \sin 2\pi(nf_0)t$$

(f)

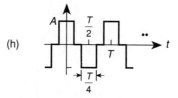

$$v(t) = \frac{2A}{\pi} \cos 2\pi f_0 t - \frac{2A}{3\pi} \cos 2\pi (3f_0)t + \frac{2A}{5\pi} \cos 2\pi (5f_0)t + \ldots$$

$$= \sum_{n=1}^{\infty} \left(A \frac{\sin n\pi/2}{n\pi/2} \right) \cos 2\pi (nf_0)t$$

(g)

$$v(t) = A\tau/T + \sum_{n=1}^{\infty} \left(2A\frac{\tau}{T} \right) \left(\frac{\sin n\pi\tau/T}{n\pi\tau/T} \right) \cos 2\pi (nf_0)t$$

(h)

$$v(t) = \sum_{n,\,\text{odd only}}^{\infty} \left(A \frac{\sin n\pi/4}{n\pi/4} \right) \cos 2\pi (nf_0)t$$

(special case of 50 per cent 'alternate inversion')

(i)

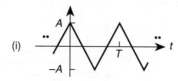

$$v(t) = \frac{8A}{\pi^2} \cos 2\pi f_0 t + \frac{8A}{9\pi^2} \cos 2\pi (3f_0)t + \frac{8A}{25\pi^2} \cos 2\pi (5f_0)t + \ldots$$

$$= \sum_{n\,\text{odd}}^{\infty} \frac{8A}{(n\pi)^2} \cos 2\pi (nf_0)t$$

(j)

$$v(t) = \frac{2A}{\pi} \left[\sin 2\pi f_0 t - \tfrac{1}{2}\sin 2\pi(2f_0)t + \tfrac{1}{3}\sin 2\pi(3f_0)t + \ldots\right]$$

$$= \sum_{n=1}^{\infty} [(-1)^{n+1}]\left(\frac{2A}{n\pi}\right)\sin 2\pi(nf_0)t$$

6

Optical fibre technology

6.1 Introduction

The transmission of information along optical fibres has been envisaged since the 1950s and tremendous advances have been made since that time. Currently the entire electrical network system is becoming a single system with broadcasting, telecommunications and computer data communications envisaged as a global system in the next millennium.

Traditional technology, such as copper cable and microwave transmission, has limitations that are overcome by fibre optics. Some of the advantages of fibre-optic cables are as follows:

- optical fibres are non-conductive – grounding and surge suppression not required;
- immune to EM energy – no radiated energy – unauthorized tapping difficult;
- large bandwidth – 50 GHz for 1 km length;
- low loss – long unrepeated links – inexpensive light sources available;
- small, lightweight cables – easy installation;
- universal use.

A typical information transmission system is shown in Fig. 6.1. This is typical of the set-up for communication applications such as the following:

- cable television
- office interlinking
- car subsystems
- monitoring

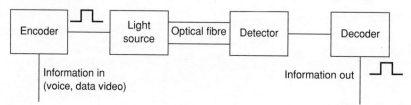

Figure 6.1

- submarine cables
- control systems
- local area networks
- telephone trunk lines
- long-distance radio links.

6.2 Fundamental principles

Light

When considering the transmission of data, voice and video it should be appreciated that many parts of the EM spectrum are used for communication purposes. Currently the visible and infrared ranges of the spectrum are being used for optical fibre transmission. The EM spectrum is shown in Fig. 6.2.

We usually think of light as a wave and an electron as a particle. However, modern physics has shown us that there is no clear distinction between these two ideas. A particle of light is called a **photon** and this can be described as a packet of wave energy. It is this concept which is generally used when discussing optical communications, although it should be appreciated that particle physics is involved in optical transmitters and receivers while wave physics is generally considered inside the fibre.

Reflection and refraction

Fundamental physics shows that when white light is passed through a prism the colours making up this light are spread out or dispersed in the prism. This is shown in Fig. 6.3. This is called **dispersion** and is met with in optical fibres, as will be seen later. The bending of the rays is caused by **refraction** and this occurs when EM waves pass from a less dense into a more dense medium, such as air to glass. Consider Fig. 6.4. In this diagram an incident ray of light is passing from a region with refractive index n_1 to a region with refractive index n_2. The definition of **refractive index** is

$$n = \frac{c}{v} \tag{6.1}$$

where c is the velocity of light in free space and v the velocity of light in a specific material. Generally the refractive index is related to a law called **Snell's law** which is expressed as

$$n_1 \sin \theta_1 = n_2 \sin \theta_2$$

It should be remembered that this law applies when light travels from a less dense into a more dense medium. However, if we consider the case where light travels from a more dense into a less dense region then a sequence of incident angle changes cause an effect known as **total internal reflection**. This is shown in Fig. 6.5. The **critical angle** at which this occurs is given by

$$\theta_c = \arcsin \frac{n_2}{n_1} \tag{6.2}$$

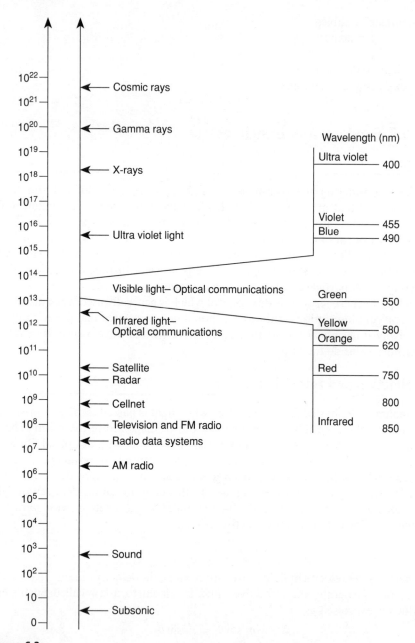

Figure 6.2

If n_2 is the refractive index of air then the expression becomes

$$\theta_c = \arcsin\frac{1}{n_1} \tag{6.3}$$

Even when light passes from one index to another, a small portion is always reflected back into the first material. These reflections are known as **Fresnel**

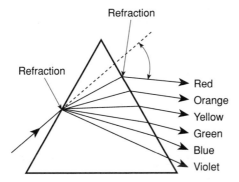

Figure 6.3

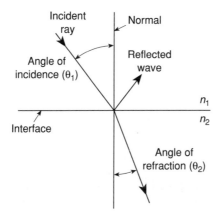

Figure 6.4

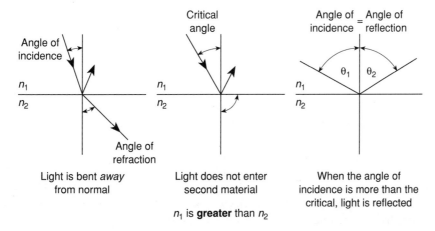

Figure 6.5

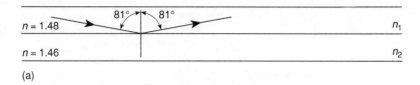

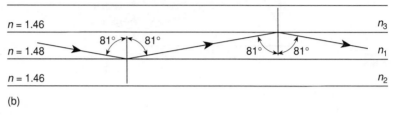

Figure 6.6

reflections. A greater difference in the indices of the materials results in a greater portion of the light reflecting. Fresnel reflection is given by

$$P = \left(\frac{n-1}{n+1}\right)^2 \tag{6.4}$$

Figure 6.6 shows an example of where total internal reflection can be used in optical fibre transmission. In Fig. 6.6(a) typical refractive index values are chosen which are used in fibre material. In Fig. 6.6(b) this example is taken a step further by adding a third layer of glass with refractive index n_3. As a result of total internal reflection the light is guided along the sandwiched layer with refractive index n_1, and as long as the angle of incidence is about 81° the light will be trapped between the two other layers. This is the principle of the optical fibre which will now be looked at in greater detail. In a practical fibre n_2 and n_3 must have a lower refractive index for total internal reflection to occur.

Modes

In order to understand the idea of modes in an optical fibre a useful concept is the formation of waves in a cavity which has restraining boundaries at either end. These principles manifest themselves in cable termination, laser cavities and microwave tubes.

Figure 6.7 shows an incident wave A and a reflected wave B interfering with one another to produce resultant waves at different moments in time. At time t_1 the phase difference is such as to produce a resultant wave R which has a lower amplitude. At time t_2 the waves are 180° out of phase and interfere destructively. At time t_3 the waves interfere constructively to produce a large resultant wave.

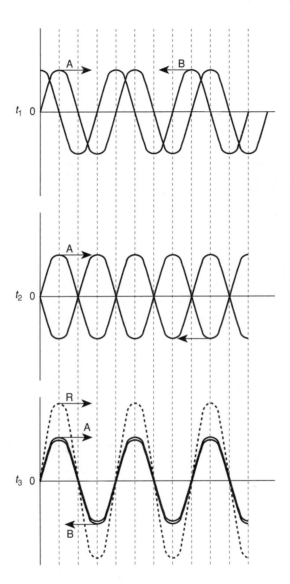

Figure 6.7

The wave pattern produced at time t_2 is called a standing wave pattern and at certain points the field is zero, while at others it oscillates within the stationary envelope. To produce a stationary standing wave pattern the cavity length (L) must be an integral number of half-wavelengths:

$$L = \frac{m\lambda}{2} \tag{6.5}$$

Only wavelengths satisfying equation (6.5) can exist inside the cavity. Waves of other lengths launched into the cavity interfere destructively.

According to equation (6.5) cavities are resonant at a number of wavelengths or frequencies and these frequencies are found by combining equations (6.1) and (6.5):

$$v = \frac{c}{n} = \lambda f$$

$$\frac{c}{n} = \frac{2Lf}{m}$$
(6.6)

$$\therefore f = \frac{cm}{2Ln}$$

The various resonant frequencies given by equation (6.6) are called the modes of the cavity. If these principles are extended to an optical fibre waveguide as in Fig. 6.6, the modes follow a zigzag path, but only certain modes will be acceptable, depending on the numerical aperture (NA) and diameter of the fibre.

6.3 Optical fibres

An optical fibre cable has two concentric layers, in its fundamental form, called the core and the **cladding**. Figure 6.8 shows the construction. Note that the cladding has a lower index of refraction than the core. Some of the light also travels into the cladding, but the cladding is usually inefficient as a light carrier and the light rays become attenuated.

Fibres have very small diameters. Some diameters of core and cladding are as follows:

Core (µm)	Cladding (µm)
8	125
50	125
62.5	125
100	140

Fibre sizes are usually expressed by first giving the core size and then the cladding size. Hence a 50/125 fibre means a core of 50 µm and a cladding of 125 µm.

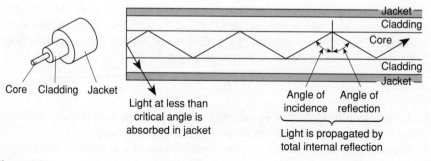

Figure 6.8

Fibre classification

Optical fibres are classified in two ways, namely by the material they are made from and second by the refractive index of the core and the modes that the fibre propagates. Optical fibres are of three basic types:

- Glass fibres in which impurities are added to vary the refractive index, e.g. germanium or phosphorus increase the index while boron and fluorine decrease it – they are generally made from silicon dioxide or fused quartz.
- Plastic-clad silica fibres (PCS) which have a glass core and plastic cladding; performance is not as good as glass fibres.
- Plastic fibres which have a plastic core and plastic cladding – they are limited in loss and bandwidth but are cheap.

The second way to classify fibres is by the refractive index of the core and the modes that the fibre propagates. Figure 6.9 shows the effects of the three types of fibre propagation. First, this figure shows the difference between the input pulse injected into the fibre and the output pulse. The broadening in width limits the fibres bandwidth. Second, it shows the path followed by the rays. Third, it shows the relative index of refraction of the core and cladding for each type of fibre.

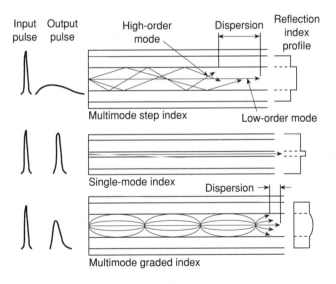

Figure 6.9

The paths followed by the rays are called **modes**. A mode is a mathematical and physical concept derived from electromagnetic theory, but for the purpose of optical transmission it will be considered here as a path that a ray can follow in travelling down a fibre. The number of modes can range from 1 to over 10 000.

Types of fibre

The refractive index profile describes the relation between the indices of the core and cladding and by this classification there are three types of fibres.

- multimode step-index fibre
- multimode graded-index fibre
- single-mode step-index fibre.

Step-index fibre

This is the simplest type and the most wide ranging, but not the most efficient in having high bandwidth and low losses.

As can be seen from Fig. 6.9, many modes are involved and consequently different path lengths result. This causes time delays and spreading due to dispersion. This is called **modal** dispersion and is similar to the spreading of wavelengths when white light passes through a prism. The effect of this is shown in Fig. 6.10. Typical modal dispersion figures for step-index fibres are 15 to 30 ns/km (nanoseconds per kilometre length). Pulse spreading results from this and limits the bandwidth as well as causing information loss.

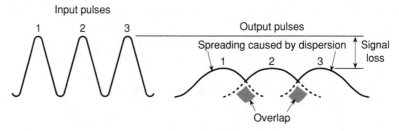

Figure 6.10

Graded-index fibre

One way to reduce modal dispersion is to use graded-index fibres in which the core has numerous concentric layers of glass, each layer having a lower index of refraction. Light travels faster in areas of lower refractive index, hence the further the light is from the centre axis, the greater the speed. Each layer of the core refracts the light and the light is now bent or continually refracted in an almost sinusoidal pattern. Those rays that follow the light travelling near the centre of the core has the slowest average velocity. As a result all rays tend to reach the end of the fibre at the same time.

These fibres are useful where wide bandwidth is required as in telecommunications networks, local area networks and computer link-ups.

Single-mode fibres

Another way to reduce modal dispersion is to reduce the core's diameter until the fibre propagates only one mode efficiently. Generally the cladding must be about 10 times thicker than the core to promote easier handling and standardize sizes. The point at which a single-mode fibre propagates only one mode depends on the wavelength of light carried. A wavelength of 820 nm results in multimode operation. As the wavelength is increased, the fibre carries fewer and fewer modes until only one remains. Single-mode operation begins when the wavelength approaches the core diameter, e.g. at 1300 nm.

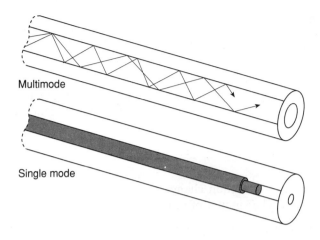

Figure 6.11

Different fibres have a specific wavelength called the **cut-off wavelength**, above which they carry only one mode. In this type of fibre some of the optical energy of the mode travels in the cladding as shown in Fig. 6.11. The diameter of the light appearing at the end of the fibre is larger than the core diameter, and this diameter is referred to as the **mode field diameter**. Although optical energy is confined to the core in a multimode fibre, it is not so confined in a single-mode diameter. The cladding in a single-mode fibre must obviously be a more efficient carrier of energy.

A final point is that a single-mode fibre can be constructed with a shorter cut-off wavelength. Some fibres have been designed with a cut-off wavelength of 570 nm for operation at 633 nm. The core is quite small, less than 4 μm. Another fibre has a cut-off wavelength of 1000 nm and a core diameter of 6 μm and this has applications in telecommunications, computer and sensor networks.

Plastic fibres

Plastic fibres have a relatively large core and very thin cladding. They are considerably less expensive than glass fibres. They also use red light in the 660 nm range and this aids troubleshooting since the presence of light in the system is easy to see. The optical power does not pose a problem of safety. They are normally easier to use and connect than glass fibres.

6.4 Fibre characteristics

This section will deal with the characteristics of optical fibres which are important to users and designers. The characteristics under consideration will be:

- dispersion
- bandwidth
- attenuation
- equilibrium mode distribution
- numerical aperture.

Dispersion

The two main types of dispersion are **modal dispersion** and **material dispersion**. Modal dispersion has already been discussed but material dispersion is of particular concern in single-mode systems.

Different wavelengths travel at different velocities through a fibre, even in the same mode. Simple physics also shows that the refractive indices, the wavelengths and the velocities are related as follows:

$$\frac{n_2}{n_1} = \frac{v_1}{v_2} = \frac{\lambda_1}{\lambda_2} \tag{6.7}$$

Hence the index of refraction changes according to the wavelength and this causes dispersion. The amount of dispersion depends on two factors:

1. The range of light wavelengths injected into the fibre. A source does not normally emit a single wavelength. The range of wavelengths for a laser is about 2 or 3 nm while an LED has a spectral width of 35 nm.
2. The centre operating wavelength of the source. Around 850 nm, longer wavelengths travel faster than shorter ones. An 860 nm wave travels through glass faster than an 850 nm wave. At 1550 nm, however, the situation is reversed. The shorter wavelengths travel faster than the longer ones, a 1560 nm wave travels slower than a 1540 nm wave. At some point, the cross-over must occur where the wavelengths travel at the same speed. This cross-over occurs around 1300 nm. This is called the **zero-dispersion wavelength**.

Figure 6.12 shows dispersion for a typical single-mode fibre. At 1300 nm dispersion is zero. At wavelengths below 1300 nm dispersion is negative so wavelengths arrive later, while above 1300 nm wavelengths arrive faster.

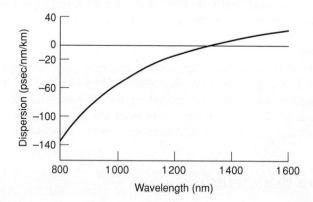

Figure 6.12

Number of modes

It has already been indicated that the number of modes supported by a fibre is partly determined by its information-carrying capacity. Modal dispersion, which causes pulse spreading and overlapping, limits the data rate that a fibre can support.

The V number, or normalized frequency, is a fibre parameter that takes into account the core diameter, wavelength propagated and the fibre numerical aperture (NA) which will be discussed later. This V number is given as

$$V = \frac{\pi d}{\lambda} \text{NA} \qquad (6.8)$$

From this number the number of modes can be determined as follows:

$$N = \frac{V^2}{2} \quad \text{(simple step index)} \qquad (6.9)$$

$$N = \frac{V^2}{4} \quad \text{(graded index)} \qquad (6.10)$$

Note that the relative refractive index Δ between core and cladding is defined as

$$\Delta = \frac{n_1 - n_2}{n_1} \quad \text{if } \Delta \leq 1 \qquad (6.11)$$

Bandwidth

Many fibre and cable manufacturers do not specify dispersion for their multimode cables. Instead, a figure of merit called the **bandwidth–length** product or simply bandwidth given in megahertz-kilometres is specified. A bandwidth of 500 kHz km means that a 500 MHz signal can be transmitted for 1 km or that the product of the frequency and the length must be 500 or less. So you can send a lower frequency a longer distance or a higher frequency a shorter distance.

Single-mode fibres, however, are specified by dispersion. This dispersion is expressed in ps/km/nm. An approximation of bandwidth can be determined from the following expression:

$$\text{BW} = \frac{0.187}{(\text{Disp})(\text{SW})(L)} \qquad (6.12)$$

where Disp is the dispersion, SW the spectral width (nm) and L the fibre length (km).

Attenuation

Attenuation is the loss of optical power as light travels along the cable and is measured in decibels per kilometre (dB/km). It is due to three basic causes:

- absorption
- scattering
- microbends.

As can be seen from Fig. 6.13, there are three basic windows involved in optical fibre transmission. Note from the graph that windows appear at 820–850 nm, the zero dispersion wavelength 1300 nm and the 1550 nm region. This is for a multimode fibre, but the loss curve for a single-mode fibre is shown in Fig. 6.14. Plastic fibres are best operated in the visible light area around 650 nm.

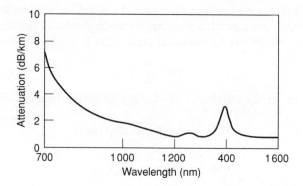

Figure 6.13

Scattering is the loss of optical energy due to imperfections in the fibre and from the basic structure of the fibre. Absorption is the process by which impurities in the fibre absorb optical energy and dissipate it as a small amount of heat. Modern manufacturing techniques reduce this considerably. Microbend loss is that loss resulting from microbends which can occur during manufacture or can be caused by the cable itself.

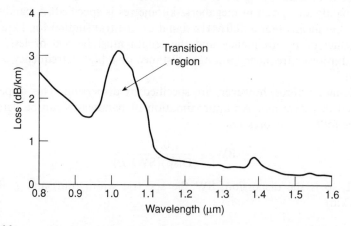

Figure 6.14

Equilibrium mode distribution

While many modes or paths are available to carry light, not all carry the same amount of light, nor do all carry light efficiently. Some modes carry no light, meaning that no energy travels along a potential path. Energy can also change paths. In a real fibre, energy transfer between modes is caused by bends in the fibre, variations in the diameter or refractive index of the core.

Over a distance light will transfer between modes until it arrives at **equilibrium mode distribution** (EMD). At this point further transfer of energy between modes does not occur under normal circumstances. It can occur under unusual circumstances such as flaws in the fibre, etc.

Before it reaches EMD, a fibre is said to be overfilled or underfilled. An overfilled fibre is one in which marginal modes carry optical energy. This energy will be attenuated or lost over a short distance. This is excess energy since for many applications it is unimportant. Some light sources, notably LEDs, can overfill a fibre. This means they inject light into modes that the fibre will not carry efficiently. Some of these modes are in the cladding. Others are high-order modes in the core that will not propagate efficiently. An underfilled fibre is one in which the light injected into the fibre fills only some of the low-order modes available for propagation of optical energy. A laser will cause this effect.

As well as the source, the type of fibre determines the distance required to reach EMD. A plastic fibre requires only a few metres or less to reach EMD. A high-quality glass fibre can require tens of kilometres before it reaches EMD.

It is important to understand EMD as the power at the end of the fibre may be different if EMD has not been reached. To compare accurately two fibres, two light sources or two connectors the conditions under which they were tested must be known, i.e. tested under EMD conditions or before EMD was reached.

Numerical aperture

Numerical aperture (NA) is the light-gathering ability of the fibre. Only light injected into the fibre at angles greater than the critical angle will be propagated. The NA is given as follows:

$$NA = \sqrt{n_1^2 - n_2^2} \tag{6.13}$$

The numerical aperture may also be given by half-angle of the cone of acceptance. This is shown in Fig. 6.15. Hence

$$\theta = \arcsin(NA) \tag{6.14}$$

A fibre with a large NA accepts light well, while a fibre with a low NA requires highly directional light. In general, fibres with a high bandwidth have a lower NA.

Generally, manufacturers do not specify NA for single-mode fibres because NA is not a critical parameter in this case. Also sources and detectors have an NA and these must be matched to the fibre's NA. This will be discussed later.

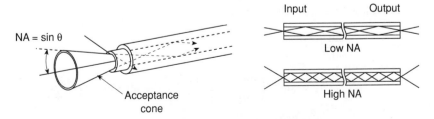

Figure 6.15

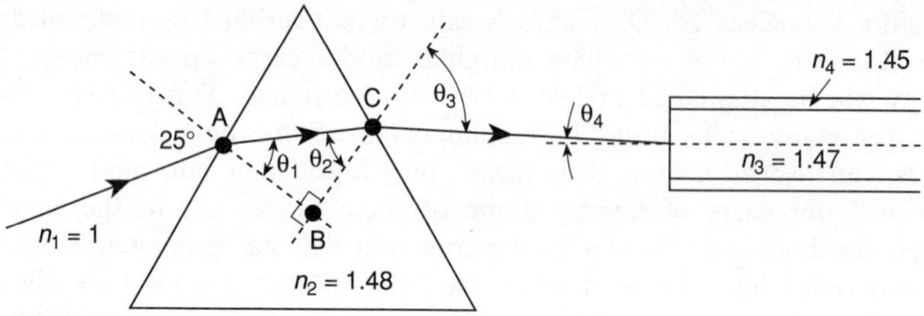

Figure 6.16

Example 6.1

A prism has to be used to direct red light at 850 nm into a fibre as shown in Fig. 6.16. Determine:

(a) the path of the light through the prism;
(b) the wavelength of the light inside the prism;
(c) the launch angle into the fibre;
(d) the critical angle inside the fibre.

Solution

(a) Using Snell's law

$$n_1 \sin 25° = n_2 \sin \theta_1$$

$$\therefore \quad \sin \theta_1 = \frac{n_1}{n_2} \sin 25°$$

$$\therefore \quad \theta_1 = \sin^{-1}\left(\frac{n_1}{n_2} \sin 25°\right)$$

$$= \sin^{-1}\left(\frac{1}{1.48} \sin 25°\right) = 16.59°$$

From the geometry of triangle ABC,

$$\theta_2 = (\theta_1 + 90°) - 180 = (16.59° + 90°) - 180 = 73.41$$

The light is now travelling from glass into air and as Snell's law only applies from a less to a more dense medium

$$\frac{\sin \theta_3}{\sin \theta_2} = \frac{1}{n_2/n_1} = \frac{1}{n_2}$$

$$\therefore \quad \theta_3 = \sin^{-1}\left(\frac{\sin \theta_2}{n_2}\right) = \frac{\sin 73.41}{1.48} = 40.35°$$

$$\therefore \quad \theta_4 = 90 - 40.35° = 49.65°$$

(b)
$$\frac{\lambda_1}{\lambda_2} = \frac{n_2}{n_1}$$

$$\therefore \lambda_2 = \frac{\lambda_1 n_1}{n_2} = \frac{850 \times 1}{1.48} = 574.32 \, \text{nm}$$

(c)
$$\text{NA} = \sqrt{n_3^2 - n_4^2} = \sqrt{(1.47)^2 - (1.45)^2} = 0.24$$

$$\therefore \theta = \sin^{-1} 0.24 = 13.98°$$

(d)
$$\sin \theta_c = \frac{1.45}{1.47}$$

$$\therefore \theta_c = \sin^{-1}\left(\frac{1.45}{1.47}\right) = 80.53°$$

Example 6.2

A single mode fibre has to operate at 850 nm and has a dispersion of 6 ps/km per nm. The light source has a spectral width of 35 nm and the length of fibre has to be 2 km. Determine:

(a) the bandwidth of the fibre under these conditions;
(b) the NA and launch angle of the fibre if the critical angle is 84° for single-mode operation and the core refractive index (n_2) is 1.49.

Solution

(a)
$$\text{BW} = \frac{0.187}{(\text{Disp})(\text{SW})(L)} = \frac{0.187}{6 \times 35 \times 2 \times 10^{-12}}$$

$$= \frac{0.187 \times 10^{12}}{6 \times 35 \times 2} = 445 \, \text{MHz}$$

(b)
$$\sin \theta_c = \frac{n^1}{n^2}$$

$$\therefore n_1 = n_2 \sin \theta_c = 1.49 \sin 84° = 1.48$$

$$\text{NA} = \sin \theta = \sqrt{n_2^2 - n_1^2}$$

$$\therefore \text{NA} = \sqrt{(1.49)^2 - (1.48)^2} = 0.17$$

$$\therefore \theta = \sin^{-1} 0.17 = 9.92°$$

Example 6.3

A multimode step index fibre with a core diameter of 62.5 μm and a relative refractive index of 1.5 per cent operates at a wavelength of 850 nm. If the core refractive index is 1.48, estimate:

(a) the numerical aperture of the fibre;
(b) the number of guided modes.

Solution

(a)
$$V = \frac{2\pi d\, \text{NA}}{\lambda}$$

for a step index fibre, also

$$\Delta = \frac{n_1 - n_2}{n_1}$$

$$\therefore\ n_2 = n_1 - n_1\Delta = 1.48 - \frac{1.48 \times 1.5}{100} = 1.46$$

$$\therefore\ \text{NA} = \sqrt{n_1^2 - n_2^2} = \sqrt{1.48^2 - 1.46^2}$$

$$= \sqrt{2.19 - 2.13} = 0.24$$

(b)
$$\therefore\ V = \frac{3.14 \times 62.5 \times 0.24 \times 10^9}{10^6 \times 850}$$

$$= \frac{3.14 \times 62.5 \times 240}{85} = 55.4$$

$$\therefore\ N = \frac{V^2}{2} = \frac{(55.4)^2}{2} = 1534$$

Example 6.4

A multimode step index fibre has a relative refractive index of 1.2 per cent and a core refractive index of 1.5. It operates at a wavelength of 1300 nm and propagates 900 modes. Estimate the diameter of the fibre core.

Solution
Since $N = v^2/2$ for simple graded index fibre

$$V = \sqrt{2N} = \sqrt{1800} = 42.43$$

Also

$$\Delta = \frac{n_1 - n_2}{n_1}$$

$$\therefore\ n_1\Delta = n_1 - n_2$$

$$0.012 \times 1.5 = 1.5 - n_2$$

$$\therefore\ n_2 = 1.48$$

$$\text{NA} = \sqrt{n_1^2 - n_2^2} = \sqrt{1.5^2 - 1.48^2} = 0.25$$

Now

$$V = \frac{\pi d}{\lambda}\text{NA}$$

$$\therefore\ d = \frac{\lambda V}{\pi\,\text{NA}} = \frac{1300 \times 42.43}{3.14 \times 0.25} = 70.26\,\mu\text{m}$$

6.5 Optical sources

The conversion of electrical energy into optical energy is achieved by the use of an optoelectronic semiconductor device such as a light-emitting diode (LED) or a laser diode. The main requirements for lasers and LED devices are:

- they should be small;
- they should be rugged;
- possess a long operating life;
- have a high coupling efficiency between semiconductor and optical fibre;
- be capable of simple modulation by a transmission signal;
- have low power consumption;
- have a large radiating power.

Light-emitting diodes

The LED meets some of the above requirements, but not all of them. They are p–n junctions that, when forward biased, will cause minority carriers to be injected across the junction. This is shown in Fig. 6.17. Once across the junction the minority carriers will recombine with the majority carriers and will give up their energy in the form of light photons.

LEDs used in fibre optics are more complex than the simple device described above because of the characteristics of wavelength and emission pattern.

The LEDs described above are called **homojunction** types, i.e. they are made from a single piece of semiconductor material. These LEDs give out large patterns with low radiance and hence most of the light emitted does not enter the fibre core. A **heterojunction** structure solves this problem by confining the carriers to the active area of the chip. This is done by using materials with different energy levels and refractive indices. These LEDs can be surface emitting or edge emitting, but the edge-emitting types give a narrower beamwidth.

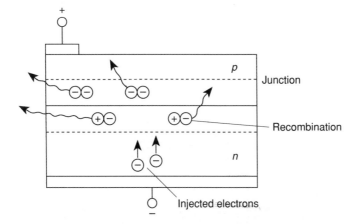

Figure 6.17

The materials used in the LED influence the wavelengths emitted. LEDs emitting in the first window of 820–850 nm are usually gallium aluminium arsenide (GaAlAs), while long-wavelength devices for use at 1300 nm are made of gallium indium arsenide phosphate (GaInAsP).

Lasers

A laser requires three conditions before lasing action takes place, namely:

- stimulated emission
- population inversion
- optical feedback.

Stimulated emission

Bohr's theory shows that emissions occur when electron transitions occur between different levels. The emission process may occur in two distinct ways, namely **spontaneous emission** and **stimulated emission**. Spontaneous emission is radiated randomly and hence is not coherent, while with stimulated emission the electron is triggered to undergo a transition by the presence of photons of energy. If a beam of light has to be amplified then the stimulated emission process must be increased in relation to the spontaneous process.

Figure 6.18 shows the basic structure of a laser diode. The main difference between this and an LED is that the laser diode has an optical cavity called a **Fabry–Pérot** cavity. They are made basically of an n-type gallium arsenide (GaAs) wafer into which p-type dopant is diffused. In the cutting process from wafer to chip form the two narrow ends are sliced to produce very smooth mirror-like surfaces called **facets** or **cavity mirrors**, while the two larger sides are deliberately roughened in the cutting process to discourage light emission.

Some of the photons emitted by the spontaneous action are trapped in the Fabry–Pérot cavity. These photons have an energy level equal to the band gap of the laser materials. If one of these photons influences an excited electron, the electron immediately recombines and gives off a photon. Since the wavelength of a photon is a measure of its wavelength, and since the energy of the

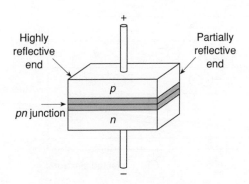

Figure 6.18

stimulated photon is equal to the original stimulating photon, its wavelength is equal to that of the original stimulating photon. The photon created is a duplicate of the first photon and has the same wavelength, phase and direction of travel. Hence amplification has occurred and emitted photons have stimulated further emission.

Population inversion

This is the process by which a larger than normal population of carriers exists at a certain energy level. When population inversion occurs a photon is more likely to stimulate emission than be absorbed.

Optical feedback

Once population inversion has been achieved the photons have to receive enough energy to leave the system, and this is the third requirement for laser action. The structure shown in Fig. 6.19 indicates that the ends are highly polished to act like mirrors. The photons are reflected back and forth until they have sufficient energy to be ejected from one of the facets. This is optical feedback.

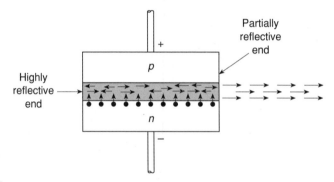

Figure 6.19

The laser thus differs from an LED in the following ways:

- Nearly monochromatic: the light emitted has a narrow band of wavelengths.
- Coherent: the light wavelengths are in phase.
- Highly directional: the light is emitted in a highly directional pattern with little divergence.

6.6 Source characteristics

The characteristics generally considered are:

- output power
- output pattern
- spectral width
- speed.

The output power of an LED is generally lower than a laser and in general the output power of the device is in the following decreasing order: laser, edge-emitting LED and surface-emitting LED. Above the threshold current an LED will have an output power of a few milliwatts, while a laser will have an output power of about 100 mW.

As light leaves a source it spreads out, producing an emission pattern. Only a portion of the light actually couples into the fibre. The output pattern of a laser is narrower than that of an LED. A good source should have a small emission diameter and a small NA and both these parameters should be compatible with the fibre parameters, e.g. a single-mode fibre requires a laser source.

Lasers and most LEDs emit a range of wavelengths and this is known as the spectral width. It is measured at 50 per cent of the maximum amplitude of the peak wavelength, e.g. if a source has a peak wavelength of 820 nm and a spectral width of 30 nm its output ranges from 805 to 835 nm. A comparison of spectral widths for lasers and LEDs is shown in Fig. 6.20.

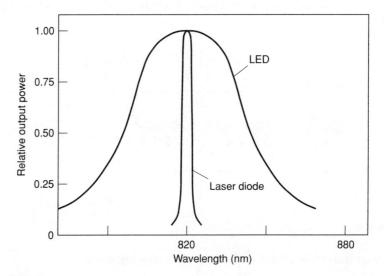

Figure 6.20

Generally spectral width is not important in optical fibre links running under 100 MHz for only a few kilometres, but it is important in high-speed, long-distance, single-mode systems because the resulting dispersion is the limiting factor.

Any source must turn on and off at a fast enough rate to meet the bandwidth requirements of the system. Source speed is specified by rise and fall times. A rough approximation is given by the following expression:

$$BW = \frac{0.35}{t_r} \qquad (6.15)$$

Note the rise time is given in nanoseconds.

6.7 Optical receivers

When considering light detectors there are three basic facts which should be remembered:

1. All light receivers are reverse-biased diodes.
2. The current induced by the photodiode is proportional to the photon energy that causes it.
3. In moving from the conduction band to the valence band a photon is emitted due to recombination. This is the basis of source operation. On the other hand in moving from the valence to the conduction band an electron is emitted. This is the case with optical detectors.

The most common light detectors are:

- the photodiode
- the PIN (p-type intrinsic n-type) photodiode
- the avalanche photodiode.

Photodiode

The operation of the photodiode may be explained by referring to Fig. 6.21. As can be seen, when a photon with energy equal to the energy gap enters the depletion region it is absorbed. As a result electron–hole pairs are formed. The holes and electrons are separated and swept through the depletion region and a current will be produced. If the hole–electron pair is produced outside the depletion region then it will recombine and no current will be produced. Hence the p-region should be as thin as possible to reduce the probability of recombination and the depletion region should be as thick as possible to improve sensitivity. Another advantage of a thick depletion region is that it reduces junction capacitance and permits faster response times.

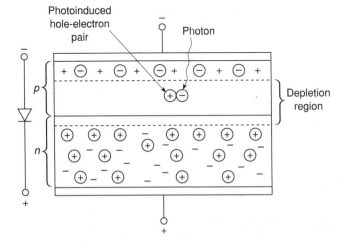

Figure 6.21

From Fig. 6.20 the p–n junction may be thought of as a capacitor with the depletion region as the dielectric. It can therefore be seen that a hole and electron created in the depletion region would be rapidly pulled apart by the oppositely charged plates. Thus the effectiveness of a p–n junction diode as a photodetector can be enhanced by widening the depletion region to give more opportunity for hole–electron pairs to form there.

PIN photodiode

The photodiode previously discussed is unsuitable for optical fibre applications because the depletion region is too narrow and also because its response is too slow. One way to increase the width of the depletion region is to include a layer of pure semiconductor material (intrinsic material) between the p- and n-type materials. Since the intrinsic layer has no free carriers, its resistance is high and the electrical forces are strong within it. The large intrinsic layer, however, means that most of the photons are absorbed within the depletion region for better efficiency. Generally, design involves balancing the twin requirements of efficiency and speed, hence the thickness of the depletion region is a compromise between these two things.

Avalanche photodiode (APD)

A photon with enough energy generates a hole–electron pair in the depletion region. The hole and electron are then swept across the depletion region by the electric field. The same principle is applied to the avalanche diode, but the effect is amplified by applying a strong reverse-bias voltage across the device.

An electron moving in the high electric field is accelerated and collides with other electrons producing further electron–hole pairs. The process is cumulative and an avalanche effect is produced, causing current amplification. At lower voltages the APD operates like a PIN photodiode and exhibits no internal gain; however, at high bias **photomultiplication** occurs due to secondary carriers caused by collision.

The basic structure of an APD is shown in Fig. 6.22. It can be seen from Fig. 6.2 that a wide depletion region is produced and hence the chance of photon-induced and electron-induced hole–electron pairs being produced is high. Sometimes the multiplication factor is mentioned in manufacturers' data sheets. This is a measure of the internal gain provided by the avalanche diode. It is defined as

$$M = \frac{I}{I_i} \tag{6.16}$$

where I is the total output current and I_i is the incident current.

Detector characteristics

The characteristics of interest are:

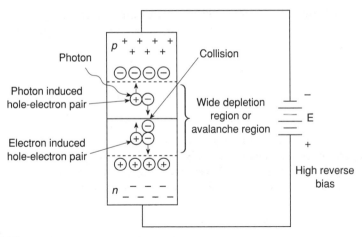

Figure 6.22

(a) responsivity
(b) quantum efficiency
(c) dark current
(d) minimum detectable power
(e) response time.

(a) The responsivity is the ratio of the diode's output current to input optical power. It is given in amps/watt. Responsivity varies with wavelength, so it is specified either at the wavelength of maximum responsivity or at a wavelength of interest. Figure 6.23 shows the responsivity plotted against wavelength for a selection of semiconductor materials. It can be seen that silicon photodiodes are not suitable for longer wavelengths of 1300 nm

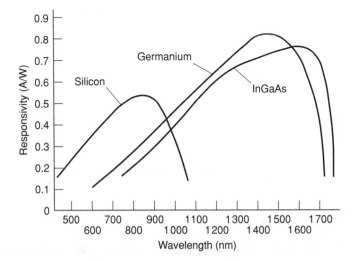

Figure 6.23

and 1550 nm. Germanium and indium gallium arsenide are more suitable for longer wavelengths. The responsivity is given by the expressions

$$R = \frac{I_o}{P_i} = \frac{\eta e}{hf} \text{ A/W} \tag{6.17}$$

where η is the quantum efficiency, I_o the output photocurrent and P_i the incident optical power.

(b) Quantum efficiency is the ratio of primary electron–hole pairs to the photons incident on the diode material. A typical quantum efficiency of 70 per cent simply means that only 7 out of every 10 photons create carriers.

(c) Dark current is the thermally generated current in a diode. It is much lower in Si photodiodes used at shorter wavelengths than in Ge and GaAs photodiodes used at longer wavelengths.

(d) The minimum power detectable by the detector determines the lowest level incident optical power that the detector can handle.

(e) Response time is the time required for the photodiode to respond to optical inputs and produce external current. It is dependent on the bias voltage: higher voltages bring faster rise times. (Response time is usually specified as a rise time and a fall time.)

The bandwidth is specified as

$$BW = \frac{0.35}{t_r}$$

This bandwidth can be limited by either the rise time or its RC time constant, hence

$$BW = \frac{1}{2\pi RC} \tag{6.18}$$

where R is the load resistance and C the diode capacitance.

Table 6.1 summarizes these parameters for the most frequently used detectors.

Table 6.1

	PIN	PIN/preamp	APD
Responsivity	80 µA/µW	2 mV/µW	70 µA/µW
Spectral response (nm)	1150–1600	1150–1600	1150–1600
Dark current (nA)	2		5
Capacitance (pF)	1.5		4
Rise time (ns)	0.5 max		0.5

6.8 Fibre-optic link

So far the components of a fibre-optic system have been discussed. In this section the actual link will be considered and this involves preliminary considerations such as whether it is advantageous to use fibres rather than conventional cables, and also whether a complete fibre-optics system should be purchased or the individual transmitters and receivers built separately.

In planning such a system the application requirements have to be specified. The main factors are type of fibre, wavelength, transmitter power, number of connectors, etc. If the power at the receiver is insufficient to meet the bit error rate (BER) requirement then there are several choices which can be used to remedy this situation, i.e. a transmitter with a higher output power can be used, the use of a fibre with a lower attenuation is possible, a shorter transmission distance may be selected or a receiver with a lower sensitivity may be used.

It can be seen, therefore, that the planning of an optical fibre system is not a simple matter and in order to assist the designer two approaches are used, called the **link power budget** and the **rise-time budget**.

Power budget

This budget analyses link losses and determines if the link can deliver enough of the transmitter's power to the receiver. It is therefore the difference between the transmitter power and the receiver power, e.g. if the peak transmitter power is $-10\,$dBm and the receiver sensitivity is $-40\,$dBm then the power budget is $30\,$dB and the link cannot tolerate more than $30\,$dB of loss. If the total loss is less than $30\,$dB, the remaining power is called the **power margin**. Typical margins are 3–6 dB.

Example 6.5

Consider a link with the following specifications in which the power margin is 6 dB:

Transmitter		**Fibre 2**	
Output power	$250\,\mu$W $(-6\,$dBm)	Size	$100/140\,\mu$m
Output diameter	$100\,\mu$m	NA	0.3
NA	0.3	Attenuation	5 dB/km
Connector loss	1 dB	Length	?

Fibre 1		**Receiver**	
Size	$85/125\,\mu$m	Sensitivity	$125\,$nA $(-39\,$dBm)
NA	0.26	Diameter	$150\,\mu$m
Attenuation	5 dB/km	NA	0.4
Length	2 km	Connector loss	1 dB
Connector loss	1 dB		

In this practical example it can be seen that the link power budget is 33 dB and the power margin is 6 dB so the power available for losses in the link is

27 dB. The other losses are now calculated in the link and the remaining loss is available for fibre 2.

Solution

Transmitter losses The transmitter losses are from NA and diameter mismatch and insertion loss from the connector which is 1 dB. Loss from diameter mismatch between the 100 µm diameter transmitter output and the 85 µm diameter fibre is

$$\text{loss}_{\text{dia}} = 10 \log_{10} \left(\frac{\text{dia}_{\text{fib}}}{\text{dia}_{\text{tr}}} \right)^2$$

$$= 10 \log_{10} \left(\frac{85}{100} \right)^2 = -1.4\,\text{dB}$$

Loss from NA mismatch between the 0.3NA of the transmitter and the 0.26NA of the fibre is

$$\text{loss} = 10 \log_{10} \left(\frac{\text{NA}_{\text{fib}}}{\text{NA}_{\text{tr}}} \right)^2$$

$$= 10 \log_{10} \left(\frac{0.26}{0.30} \right)^2 = -1.2\,\text{dB}$$

The loss at the transmitter interface is now 3.6 dB, i.e. 1 dB + 2.6 dB.

Fibre losses Since the attenuation for the 85/125 fibre is 5 dB/km, the loss for a 2 km run is 10 dB. The total loss is now 13.6 dB.

Fibre-to-fibre connection Since fibre 2 has a larger core diameter and NA than fibre 1, no mismatch losses occur. The only loss is connector insertion loss of 1 dB. The total loss is now 14.6 dB.

Receiver losses The loss at the receiver is simply the 1 dB from the connector. Since the diameter and the NA of the detector in the receiver are larger than the diameter and NA of the fibre, no mismatch occurs. The total loss is now 15.6 dB.

Fibre 2 loss The losses calculated so far total 15.6 dB. Since allowed losses were 27 dB, there is 11.4 dB available for attenuation in the 100/140 fibre. This fibre has an attenuation of 5 dB/km, so the length of fibre 2 can be 2.28 km. Figure 6.24 shows the application together with the losses and power levels at each point in the system. This is a **link power budget gradient**.

Consider the case where the length of fibre 2 is set at 0.5 km. This would leave 8.6 dB in the power budget so that the power output of the transmitter could be reduced to extend the life of the source, or the receiver could be operated above the required sensitivity to obtain a better BER.

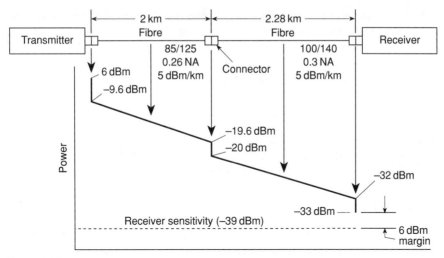

Figure 6.24

The budget may now be given as follows:

Transmitter power	−6 dBm	
Receiver sensitivity	−39 dBm	
Power budget		33 dB
Transmitter losses	3.6 dB	
Fibre 1 loss (2 km)	10 dB	
Fibre-to-fibre loss	1 dB	
Fibre 2 loss (2.28 km)	11.4 dB	
Receiver loss	1 dB	
Total loss		27 dB
Power margin		6 dB

Rise-time budget

The power budget showed that sufficient power was available for application demands. However, bandwidth is the other factor involved in the link and this is referred to as the rise-time budget. This aspect of the link deals with operational speed of the various components in the system and hence is connected with the rise and fall times of devices.

When the bandwidth of a component is specified, its rise time can be approximated from

$$t_r = \frac{0.35}{\text{BW}}$$

This equation accounts for the modal dispersion in a fibre. The rise time must be scaled to the fibre length used in the application. If the cable is specified at 600 MHz/km and the application is for a 2 km path, the bandwidth is 300 MHz and the rise time is 1.6 ns.

The rise-time budget must also include the rise times of the transmitter and receiver rather than the source and detector, because the transmitter and receiver circuits limit the maximum speed at which optoelectronic devices can operate. Connectors, splices and couplers do not affect the rise-time budget.

The system rise time can be calculated from the following:

$$t_{rsys} = 1.1\sqrt{t_{r_1}^2 + t_{r_2}^2 \ldots t_{r_n}^2} \tag{6.19}$$

Note the 1.1 allows for a 10 per cent degradation factor in the system rise time, but is not used when rearranging the formula to determine an individual rise time. This is shown below in the following example.

Example 6.6

A system operates at 20 MHz over a distance of 2 km. The fibre used has a 400 MHz-km bandwidth. If the receiver rise time is 10 ns, what is the rise time for the transmitter, assuming that the system rise time is 17.5 ns and the fibre rise time is 1.75 ns?

Solution
The transmitter rise time is given by

$$t_{rtrans} = \sqrt{17.5^2 - 10^2 - 1.75^2} = 14.25 \text{ ns}$$

The transmitter must have a rise time of about 14 ns, hence if the transmitter is selected with a rise time of 10 ns, the rise-time requirements will be met. This will give a system rise time of

$$t_{rsys} = 1.1\sqrt{10^2 + 1.75^2 + 10^2} = 15.7 \text{ ns}$$

This is within the required value of 17.5 ns.

Example 6.7

A photodiode has a quantum efficiency of 60 per cent when photons of energy 1.5×10^{-19} J are incident upon it. Calculate:

(a) the wavelength at which the photodiode operates;
(b) the incident optical power required if a photocurrent of 2.5 μA has to be obtained for satisfactory current amplification.

Solution

(a) Recall that photon energy is given as

$$E = \frac{hc}{\lambda}$$

$$\therefore \ \lambda = \frac{hc}{E} = \frac{6.62 \times 10^{-34} \times 3 \times 10^8}{1.5 \times 10^{-19}} = 1.33 \ \mu m$$

(b)
$$R = \frac{\eta e}{hf} = \frac{0.6 \times 1.6 \times 10^{-19}}{1.5 \times 10^{-19}} = 0.64 \, \text{A/W}$$

$$\therefore P_1 = \frac{I_o}{R} = \frac{2.5 \times 10^{-6}}{0.64} = 3.9 \, \mu\text{W}$$

Example 6.8

The quantum efficiency of an avalanche diode is 70 per cent when radiation having a wavelength of $0.85 \, \mu\text{m}$ has to be detected. If the optical power incident on the diode is $0.6 \, \text{mW}$, determine the output current if the multiplication factor is 30.

Solution

$$R = \frac{\eta e \lambda}{Lc} = \frac{0.7 \times 1.6 \times 10^{-19} \times 0.85 \times 10^{-6}}{6.6 \times 10^{-34} \times 3 \times 10^8} = 0.48 \, \text{A/W}$$

Also

$$I_o = RP_i = 0.6 \times 10^{-3} \times 0.48 = 288 \, \mu\text{A}$$

$$M = \frac{I}{I_i}$$

$$\therefore I = MI_i = 30 \times 288 \, \mu\text{A} \times 10^{-6} = 8.64 \, \text{mA}$$

6.9 Further problems

1. A step index fibre has an NA of 0.16, a core refractive index of 1.44 and a core diameter of $60 \, \mu\text{m}$. Determine the normalized frequency for the fibre when light at a wavelength of $900 \, \text{nm}$ is transmitted. Also estimate the number of modes in the fibre.
   Answer: 34 560

2. The velocity of light in the core of a step index fibre is $2.02 \times 10^8 \, \text{m/s}$. If the critical angle at the core–cladding interface is $80°$, determine:
   (a) the NA
   (b) the acceptance angle
   for the fibre.
   Answer: 0.26; 15.2°

3. A single-mode step-index fibre has a core diameter of $4 \, \mu\text{m}$ and a core refractive index of 1.49. Estimate the shortest wavelength of light which allows single-mode operation when the relative refractive index difference for the fibre is 1.8 per cent.
   Answer: $1.5 \, \mu\text{m}$

4. A data transmission system has a speed of $10 \, \text{Mbps}$ and uses an LED as a light source and a PIN detector. Calculate the maximum cable span if the following specification is given:

- 850 nm operation
- launch power is -16 dBm
- minimum received power is -44 dBm
- two connectors at 1.5 dB loss each
- ten fusion splices at 0.2 dB loss each
- fibre is multimode 62.5/125 µm with a loss of 3 dB/km.

Allow 6 dB margin for variation in cable, connector splices, LED power and PIN sensitivity.

Answer: 5.6 km

5. An optical fibre system uses fibre which has a loss of 7 dB/km. The average splice losses are 1.5 dB/km and the connector loss total 8 dB. If the total permitted channel loss is 37 dB, determine the maximum transmission distance possible.

Answer: 3.39 km

6. A digital single mode optical fibre system is designed to operate at 1500 nm. The transmission rate is 550 Mbps for an operating distance of 48 km. The laser launches an optical power of -14 dBm into the fibre cables which has a loss of 0.38 dB/km. If the splice losses are 0.1 dB/km, the connector loss at the receiver is 0.5 dB and the receiver sensitivity is -39 dBm, calculate the optical power budget for the system and determine the safety margin.

Answer: 2 dB

7. If an avalanche diode has a quantum efficiency of 44 per cent at an operating wavelength of 850 nm, calculate the optical power received from a fibre at this wavelength if the output photocurrent from the diode is 15 µA and the gain of the diode is 320.

Answer: 0.07 µW

8. A photodiode has a quantum efficiency of 62 per cent when operating at 1500 nm. Calculate:
   (a) its responsivity at this wavelength;
   (b) the optical power received if the mean photocurrent is 2.5 µA.

Answer: 0.75 A/W; 3.3 µW

9. A single-mode fibre has to operate at 1550 nm and has a dispersion of 8 ps/km per nm. If the light source has a spectral width of 4 nm and the length of fibre has to be 1.5 km, determine:
   (a) the bandwidth of the fibre under these conditions;
   (b) the NA and launch angle of the fibre if the critical angle is 78° for single-mode operation and the core refractive index is 1.45.

Answer: 3.89 GHz; 0.29; 17°

10. A multimode step index fibre has a relative refractive index difference of 1.6 per cent and a core refractive index of 1.47. It operates at 1550 nm and propagates 750 modes. Estimate the diameter of the fibre core.

Answer: 79.6 µm

Application of interference and polarization to antennae configurations

7.1 Introduction

Figure 7.1 shows typical satellite and terrestrial television reception equipment. Two antennae units are used, one a dish, the other a high-gain Yagi array. These systems enable the reception of low-level signals by means of the high-gain and directivity parameters of the antennae.

The principles of interference and polarization are both used in modern antenna design and both of these phenomena will be discussed in this chapter.

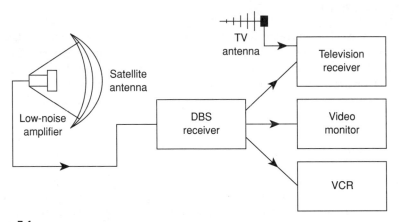

Figure 7.1

7.2 Electromagnetic propagation

Any discussion on antennae systems involves radiation in the form of an electromagnetic field. This implies that both a magnetic and electric field are present.

Electric fields

It is usual to represent electrostatic fields by lines of force as shown in Fig. 7.2. In this case two point charges are shown at A and B, separated by distance d. The lines of force act in the direction shown as does the field strength, and the potential difference is given in V volts. This concept may be extended to a capacitor which, being a parallel plate device separated by a medium, will have an electric field between its plates. The strength of this field, however, will depend, among other things, on the area of the plates and the dielectric material.

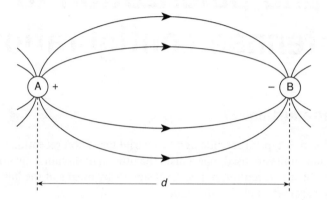

Figure 7.2

Magnetic field

When an electric current flows through a conductor, a magnetic field is produced around the conductor in the form of concentric circles. As with electrostatic fields, both magnitude and direction can be attributed to this force. However, the direction of the force is governed by the direction in which the current is flowing, and this direction can be determined by the right-hand screw thread rule.

Electromagnetic fields

The two effects of electric and magnetic energy transfer are manifested in a straight piece of wire due to a current flowing in the wire, this current being a result of the p.d. between the ends of the wire. If the current through the wire is d.c., then the polarity of the p.d. at the ends of the wire will change with direction of current and energy will radiate from the wire in the form of an electromagnetic wave in which there is an interchange of energy between the electric and magnetic fields.

As shown in Fig. 7.3(a), a conductor is connected to the positive and negative terminals of an a.c. generator. Due to the inductance of the conductor at high frequencies, a large voltage drop exists between the two ends of the wire. Hence, a strong electric field is produced as shown by the solid arrows. Figure 7.3(b)

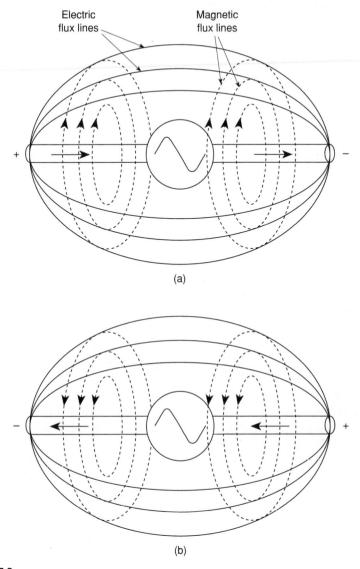

Electric flux lines

Magnetic flux lines

(a)

(b)

Figure 7.3

shows the direction of the electromagnetic field on the second half of the cycle. As the waves move outward their curvature becomes less and less, until at great distances from their origin the curvature of these distant waves is so slight that any small section appears to be a straight line. Hence plane waves are produced which travel through space in a manner shown in Fig. 7.4.

Figure 7.4 summarizes the main features of an electromagnetic wave and these features are given below:

• Energy is radiated in the form of an electromagnetic field.

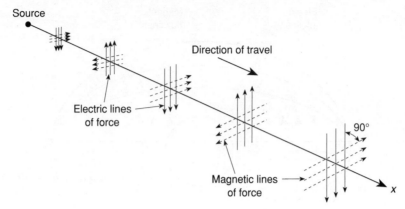

Figure 7.4

- The electric and magnetic fields vary in two planes at right angles to each other and the direction of propagation.
- The electric and magnetic field strengths are in phase except where the electromagnetic wave is at a distance of approximately 10λ.

7.3 Current and voltage distributions in antennae

In discussing the theory behind any antenna system as a radiator of energy it is informative to investigate the properties of transmission lines which carry contained energy from source to load and, except for internal losses, do not lose energy through radiation.

Each transmission line has its own characteristic impedance (Z_0) and represents this same impedance regardless of the length of the line. When the load and generator impedances are both equal to the characteristic impedance of the line maximum energy is transferred. If the line is terminated in any other resistance, standing wave patterns are set up.

A variety of conditions may be obtained by using either an open-circuited or short-circuited transmission line cut to the desired length. In those cases where it is desired to reflect a certain impedance back to the sending end, one of these two conditions is used. Four conditions of reflected impedance are possible with either open- or short-circuited lines, i.e. parallel resonance, series resonance, capacitance and inductance. These are illustrated in Fig. 7.5. Figure 7.5(f) shows an open-ended $\lambda/4$ section where the current is minimum and the voltage maximum at the termination. A $\lambda/4$ away, at the generator, reflections result in maximum current, i.e. the condition of a low-impedance circuit. In effect, the $\lambda/4$ open line reflects the equivalent of a low-impedance series resonant circuit.

If the two conductors are opened through $90°$ until they are in the same straight line, a $\lambda/2$ dipole results with the current and voltage distribution shown in Fig. 7.6.

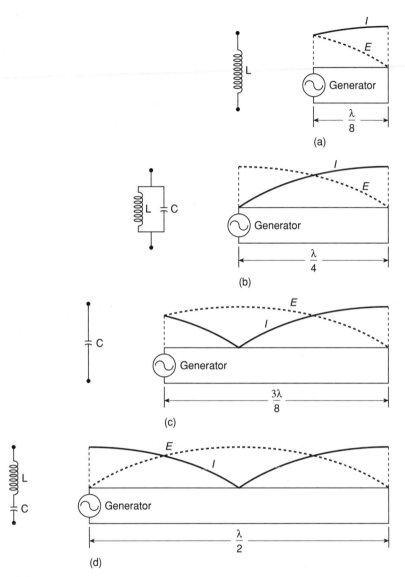

Figure 7.5a–d

7.4 Antenna parameters

Impedance of a dipole

The antenna so far developed is a conductor whose electrical length is half the wavelength at the desired frequency of oscillation and which is centre fed. If the antenna is used over a wide range of frequencies, the mean frequency is chosen.

From Fig. 7.6 it can be seen that as the voltage is large at both ends of the antenna and the current is small then the impedance is high (in the order of

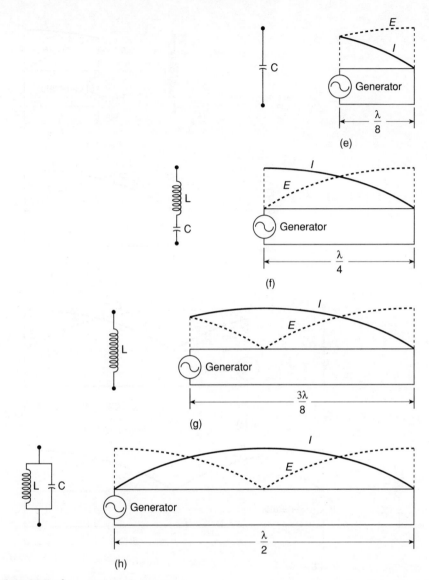

Figure 7.5e–h

$3700\,\Omega$). At the centre the current is large, the voltage is small and the dipole impedance is small (in the order of $73\,\Omega$). Hence when the impedance of an antenna is referred to it is necessary to indicate the antenna point to be considered. Usually the feed point is chosen.

Antenna physical length

It has already been mentioned that the half-wave dipole has an electrical length of half the operating frequency. The physical length is slightly less than this, i.e. 5 per cent shorter in practice.

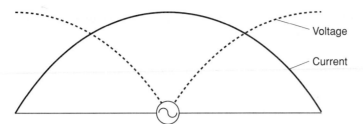

Figure 7.6

If the physical length is exactly a half-wavelength, the impedance of the dipole has an inductive characteristic, whereas if the physical length of the dipole is slightly less than half a wavelength, the input impedance is a pure resistance. Under these conditions its electrical length is half a wavelength.

It follows that a dipole of a particular length exhibits either an inductive, capacitive or a resistive characteristic at its centre, depending upon the frequency of the applied signal. When the applied frequency is higher than the resonant frequency of the dipole the length of the dipole is greater than half a wavelength. It then has the characteristic of a series-tuned circuit above resonance, i.e. inductive. Similarly, if the signal frequency is lower than resonance the antenna is capacitive.

Effective length

The effective length (or height) of a simple antenna is defined as

$$l_{\text{eff}} = \frac{E}{F} \text{ m} \tag{7.1}$$

where E is the induced voltage (V), F the field strength (V/m) and l the effective height of the antenna (m). It should be noted that the effective length of the antenna has nothing to do with the actual length of the mast on which it is mounted. It is more correctly associated with the type of antenna employed. The $\lambda/2$ dipole has an effective length given by

$$l_{\text{eff}} = \frac{\lambda}{\pi} \tag{7.2}$$

while for a folded dipole (see later) the effective length is

$$l_{\text{eff}} = \frac{2\lambda}{\pi} \tag{7.3}$$

The significance of the effective length is its relationship to radiation resistance and consequently the radiated power of an antenna.

Antenna radiation resistance

For any given reference point on the antenna, the radiation resistance (R_r) is given by

$$R_r = \frac{\text{power radiated by the antenna}}{(\text{r.m.s. current at the reference point})^2} \tag{7.4}$$

Notice that this resistance represents the function of converting electrical power into radiation, but there is no actual resistor associated with radiation resistance.

The energy fed to the antenna is dissipated by radiation and heat in the system. The energy radiated is lost the same as that dissipated as heat, in the sense that neither is returned to the circuit and must be resupplied by the transmitter to maintain the antenna current. Thus, at high frequencies, the source must supply much more energy than is dissipated as heat in the antenna, and this extra energy is dissipated in the form of radiated electromagnetic waves.

More technically, radiation resistance is defined as the ratio of the radiated power to the square of the antenna current as has already been shown, and it is a measure of how well the antenna couples the transmitter output to the surrounding space. The same argument holds for a receiver antenna. For good antenna efficiency the radiation resistance should be as high as possible compared to the heat loss resistance of the antenna.

It may also be shown that the radiation resistance and the effective length are related for a $\lambda/2$ dipole:

$$R_r = \frac{160\pi^2 l_{\text{eff}}^2}{\lambda^2} \tag{7.5}$$

hence the power radiated (P_r) is given by

$$P_r = I^2 R_r = \frac{I^2 160\pi^2 l_{\text{eff}}^2}{\lambda^2} = \frac{I^2 1580 l_{\text{eff}}^2}{\lambda^2} \tag{7.6}$$

From this it is evident that to radiate efficiently l_{eff}/λ must be as large as possible.

Example 7.1

A UHF antenna is designed to transmit a 620 MHz signal. The antenna current is measured as 1 A. Determine the radiated power of the antenna.

Solution
The effective length is given by

$$l_{\text{eff}} = \frac{\lambda}{\pi}$$

also

$$\lambda = \frac{v}{f} = \frac{3 \times 10^8}{620 \times 10^6} = \frac{30}{62}$$

$$l_{\text{eff}} = \frac{\lambda}{\pi} = \frac{30}{62 \times \pi} = 0.15\,\text{m}$$

$$P_r = \frac{1 \times 160 \times 9.86 \times 0.0225}{0.234} = 151.7\,\text{W}$$

Antenna directivity

All practical antennae radiate better in some directions than others. This characteristic is known as **directivity**. The field around a transmitting antenna is known as an **envelope**, the shape of which indicates the direction and relative strength of the radiation. The envelope is plotted by means of a suitable measuring instrument, held at different points around the antenna at distances which provide equal meter readings. For the $\lambda/2$ antenna the test points of equal readings all lie on the surface of an imaginary doughnut-shaped envelope. This is shown in Fig. 7.7.

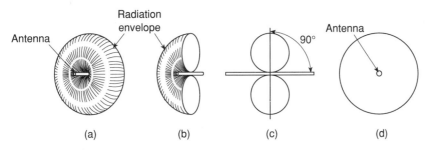

Figure 7.7

Regardless of the direction of the diameter at which it is taken, a cross-section of the radiation envelope of Fig. 7.7(a) has the form shown in Fig. 7.7(b). Therefore it is common practice to employ simplified directivity patterns like those of Fig. 7.7(c) and (d) to indicate the directional characteristics of an antenna.

The radiation field of an antenna depends upon whether the antenna is mounted vertically or horizontally. For example, Fig. 7.7(c) may represent a top view of the directivity of a horizontal $\lambda/2$ antenna. But mounted vertically the same antenna has directivity as indicated in Fig. 7.7(d) when viewed from the top.

7.5 Dipoles with secondary radiators

Dipoles can be placed near to a conducting plane to enable a gain to be obtained in a particular direction. It is not always convenient to mount a plane surface near to a dipole so an alternative is employed which consists of a single element, usually known as a secondary or parasitic radiator, used in conjunction with the single dipole. When the element is placed behind the dipole it is known as a **reflector** and when placed in front it is known as a **director**. This is shown in Fig. 7.8.

It should be remembered that an antenna intercepts a certain amount of energy from an electromagnetic field in which it is situated. If the antenna and receiver are matched then half of the energy is passed to the receiver and the other half absorbed by the antenna resistance, i.e. it is reradiated by the antenna.

As the reflector is in the shadow of the dipole it absorbs this reradiated energy. This electrical energy cannot be channelled off in the usual way because

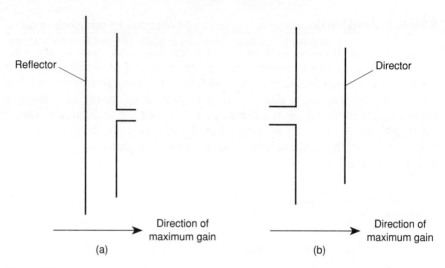

Reflector

Director

Direction of maximum gain

Direction of maximum gain

(a)

(b)

Figure 7.8

the reflector is not electrically connected to the dipole. Hence the energy absorbed by the reflector is radiated again and the dipole thus receives not only the energy which it absorbs direct from the electromagnetic field but also energy reradiated from the reflector.

Usually the dipole reflector spacing is 0.25 of the desired wavelength and the length is 5 per cent longer than the dipole. If this is adhered to, the electric field strengths of the two fields reinforce each other and a voltage is induced in the dipole which is higher than without the reflector. The following features of a dipole are affected by a reflector:

- The effective length is increased.
- The bandwidth is decreased.
- The resistance is decreased.
- The radiation pattern is altered as shown in Fig. 7.9.
- Directivity and gain are increased.

Figure 7.9

A further increase in directivity and gain can be obtained by using one or more directors. Usually a maximum of nine directors is used, as after this number no further benefit in gain is achieved due to phase differences and propagation delay. The first director is 5 per cent shorter than the dipole while the length of subsequent directors reduces progressively by 2.5 per cent.

An antenna system which uses a driven element and several parasitic elements is known as a Yagi array. Such an array, however, normally has its centre impedance reduced to a low value by use of the parasitic elements and because of this a folded dipole is used. This increases the centre impedance from 75 to 300 Ω.

Gain and directivity

From the previous section on parasitic elements it has been shown that gain and directivity are interrelated. The gain of an antenna is defined in terms of the power produced by a reference signal, the transmitted power level and receiving conditions remaining constant. The reference antenna is usually a $\lambda/2$ dipole. This means that if a receiving antenna has a gain of 3 dB it indicates that the power supplied to the terminals of a receiver by that antenna is twice that which would be supplied to the receiver if the receiver were replaced by a simple $\lambda/2$ dipole. This also applies to a transmitting antenna.

For multi-component antennae the following approximate relationship can be used:

$$l_{\text{eff}} = \frac{\lambda}{\pi} G \qquad (7.7)$$

where G is the antenna power gain in decibels. Note that this gain is roughly equal to the square root of the number of components in the antenna, so this will give the number of elements needed and also the type of antenna.

Example 7.2

A Yagi array consists of a $\lambda/2$ dipole with a reflector and director. Calculate the approximate dimensions and spacings for the elements if the operating frequency is 55 MHz.

Solution
Since the operating frequency is 55 MHz then

$$\lambda = \frac{v}{f} = \frac{3 \times 10^8}{55 \times 10^6} = 5.45 \, \text{m}$$

therefore $\lambda/2 = 2.72 \, \text{m}$.

The reflector should be 5 per cent longer and $0.25\,\lambda$ behind the dipole, therefore reflector length is

$$0.136 + 2.72 = 2.86 \, \text{m}$$

$$\text{Reflector/dipole spacing} = 0.25 \times 5.45 = 1.36 \, \text{m}$$

The director should be about 5 per cent shorter than the $\lambda/2$ dipole and 0.15 in front of the dipole:

$$\text{Director length} = 2.72 - 0.136 = 2.584 \, \text{m}$$

$$\text{Director/dipole spacing} = 0.15 \times 5.45 = 0.82 \, \text{m}$$

Example 7.3

A mobile UHF satellite receiver works in the band 602–608 MHz. Assuming the field strength is $2000 \, \mu\text{V/m}$ and the induced voltage is 1 mV, determine the number of elements in the required array.

Solution

$$l_{\text{eff}} = \frac{E}{F} \frac{10^6}{10^3 \times 2000} = 0.5 \, \text{m}$$

Average frequency is

$$\frac{602 + 608}{2} = 605 \, \text{MHz}$$

Therefore

$$\lambda = \frac{v}{f} = \frac{3 \times 10^8}{605 \times 10^6} = \frac{300}{605} = 0.5 \, \text{m}$$

Since

$$l_{\text{eff}} = \frac{\lambda}{\pi} G$$

then

$$G = \frac{l_{\text{eff}} \pi}{\lambda} = \frac{0.5 \times \pi}{0.5} = 3.14 \, \text{dB}$$

Hence $3.14 \times 3.14 = 9.85$. Hence 10 elements are required.

As has already been mentioned, a vertical dipole does not possess any directional properties in a horizontal plane as can be seen from Fig. 7.7(d). Also, when placed horizontally its directional properties are not particularly marked. This is not always a desirable feature and, in the frequency range in which half-wave dipoles are used, transmission and reception are usually required in one particular direction. It would therefore be more economical to have an antenna which radiates all the available power in one particular direction instead of radiating it in all directions.

The Yagi system has already been mentioned. Another way in which an antenna can have its directivity improved is to use a number of dipoles mounted vertically or horizontally. Depending upon the spacing between them and the way in which they are fed, the directivity of the combination can be altered. This is a more complex system, but there is a greater degree of flexibility built into such an arrangement. Furthermore, such a system can also be used in conjunction with secondary radiators which increase the directivity still further.

7.6 Principles of wave interference

In order to understand the radiation patterns and field strengths associated with multiple antennae systems an intuitive approach to wave interference is required and this is best appreciated if light sources are used, although this phenomenon occurs throughout the electromagnetic spectrum.

Interference occurs when two waves interact, but certain conditions are required if the effects of interference are to be detected or observed:

- There must be a constant phase difference between them and hence they must have the same frequency. This phase difference may be zero.
- The interfering waves must have approximately the same amplitude, otherwise the contrast or definition of the interference pattern is poor.
- When light is emitted by a source it does so because electrons jump randomly to different energy levels within the atom (the Bohr theory) and as there must be a constant phase difference between two waves they must come from the same point on the same source. If this is not done the interference pattern changes at such a speed as to give the impression of a uniform light pattern.
- Diffraction is necessary before interference will occur.

Young's double slit experiment

This experiment generally uses a monochromatic light source as shown in Fig. 7.10. S, S_1 and S_2 are narrow slits and hence diffraction takes place at each slit, thus causing an interference pattern on the screen in the form of light and dark fringes.

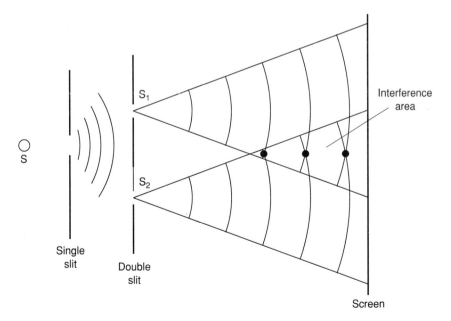

Figure 7.10

Calculation of fringe separation

Figure 7.11 shows the relative positions of the coherent sources and the geometry involved in determining the fringe pattern. The path difference $(S_2R - S_1R)$ between waves reaching R from S_1 and S_2 can be found as follows:

$$(S_2R)^2 - (S_1R)^2 = \{(S_2Y)^2 + (YR)^2\} - \{(S_1X)^2 + (XR)^2\}$$

$$= (YR)^2 - (XR)^2 = (YR + XR)(YR - XR)$$

$$= \left[\left(x_R + \frac{d}{2}\right) + \left(x_R - \frac{d}{2}\right)\right]d = 2dx_R$$

$$\therefore S_2R - S_1R = \frac{2dx_R}{S_2R + S_1R}$$

Generally $d \ll D$ so $(S_2R + S_1R)$ can be taken to be equal to $2D$ (if R is close to 0)

$$\therefore S_2R - S_1R = \frac{dx_R}{D} \tag{7.8}$$

If $(S_2R - S_1R)$ equals a whole number of wavelengths (λ) then

$$S_2R - S_1R = n\lambda \quad (n = 0, 1, 2, \ldots)$$

From equation (7.8)

$$\frac{dx_R}{D} = n\lambda$$

$$\therefore x_R = \frac{n\lambda D}{d} \tag{7.9}$$

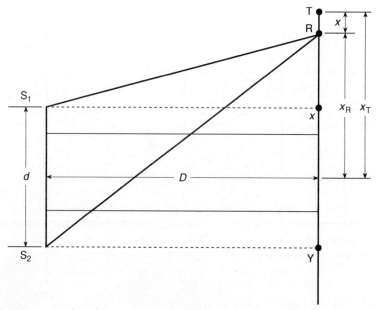

Figure 7.11

If T is the point where the next bright fringe appears then

$$x_T = (n+1)\frac{\lambda D}{d}$$

The separation of the fringes at R and T is

$$(x_T - x_R) = (n+1)\frac{\lambda D}{d} - \frac{n\lambda D}{d} = \frac{\lambda D}{d}$$

Hence dark fringes appear where the path differences are equal to odd numbers of half-wavelengths. So generally

$$x = \frac{\lambda D}{d} \tag{7.10}$$

where x is the separation of dark or bright fringes.

The Young's slits experiment offers the following conclusions which can be applied to antennae configurations:

- The fringe separation is increased by increasing D, but the fringe intensity is reduced. In a similar way antenna range is reduced.
- The fringe separation is increased by decreasing d. As will be seen in future sections in the chapter, antennae systems develop side loss as d is increased. This is analogous to a decrease in fringe separation.
- Increasing the width of any of slits S_1 and S_2 increases the intensity of the pattern, but the fringes are less sharply defined. This is analogous to altering the gain and directivity of an antennae system.

7.7 Two driven isotropic radiators

This is sometimes known as **pattern multiplication** and the resultant field pattern of an array of isotropic sources may be generally expressed in two parts as shown:

$$E_{\theta r} = F_1 + F_2 \tag{7.11}$$

where F_1 is the field pattern of an isotropic source and F_2 the array factor already mentioned and which is normally given for N radiators (in this text only two have been considered).

The expression above enables the polar diagram to be determined for arrays in which the elements may be other than isotropic elements. This will be demonstrated in the examples which follow. It will be seen that the final field pattern in a particular plane is obtained by multiplying the two fields F_1 and F_2.

To show how the fields from two antennae can combine to form a composite field, two **isotropic** radiators radiating in phase and spaced a distance d apart will be considered. An isotropic radiator is one which radiates uniformly in all directions. No such antenna exists in practice because all the radiation is assumed to come from a point source. However, it is a useful concept when considering antennae theory and can also be used as a standard radiator with which other antennae can be compared.

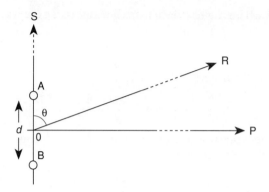

Figure 7.12

Figure 7.12 shows two isotropic radiators A and B spaced at a distance d apart. Although radiation from each is omnidirectional, only the effects of the two fields in the plane of the paper will be considered. If E is the field strength due to a given power radiated from one radiator, then if the same power is divided between A and B the field strength due to A alone is $E/\sqrt{2}$ and that due to B alone is $E/\sqrt{2}$. Let the field strengths due to A and B be E_A and E_B respectively. If a point P is considered which is a large distance from A and B and in a direction at right angles to the line joining A and B the total field strength is given by

$$E_A + E_B = E\sqrt{2}$$

The fields add because radiations from A and B have the same distance to travel and arrive at P in phase. The value of field strength at P is greater than that due to the same power radiated from one radiator. This may seem strange since the same power has only been divided between two radiators. However, this is the first indication that there is an increase in antennae gain on one particular direction.

Consider a point R in a direction θ to the line joining A and B. The field strengths due to A and B still remain at E_A and E_B respectively, but the wave from B has farther to travel to reach R so the fields will not arrive at R in phase. The vector sum of the field strengths due to A and B must therefore be calculated. From Fig. 7.13 it can be seen that the extra distance travelled

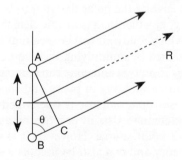

Figure 7.13

by the wave from B is BC. The wave from B will thus lag the wave from A by an angle which will be called φ. Now

$$BC = d\cos\theta$$

$$\varphi = \frac{d}{\lambda}\cos\theta \text{ wavelengths} \tag{7.12}$$

$$\therefore\ \varphi = \frac{2\pi d}{\lambda}\cos\theta \text{ radians}$$

The vector sum E_A and E_B is shown in Fig. 7.14. E_B lags E_A by an angle φ and the resultant of E_A and E_B is E_θ, where θ indicates the angle made with the line AB in Fig. 7.13. From Fig. 7.14

$$\frac{E_\theta}{2} = E_B\cos\frac{\varphi}{2}$$

$$\therefore\ E_\theta = 2E_B\cos\frac{\varphi}{2} \tag{7.13}$$

From equation (7.12)

$$\frac{\varphi}{2} = \frac{\pi d}{\lambda}\cos\theta \text{ radians} \tag{7.14}$$

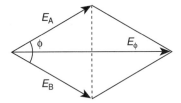

Figure 7.14

Substituting equation (7.14) in equation (7.13)

$$E_\theta = 2E_B\cos\left(\frac{\pi d}{\lambda}\cos\theta\right)$$

But

$$2E_B = \frac{E}{\sqrt{2}} + \frac{E}{\sqrt{2}} = \sqrt{2}E$$

$$\therefore\ E_\theta = \sqrt{2}E\cos\left(\frac{\pi d}{\lambda}\cos\theta\right) \tag{7.15}$$

$$= E \times \text{array factor} \tag{7.16}$$

where the 'array factor' is

$$\sqrt{2}\cos\left(\frac{\pi d}{\lambda}\cos\theta\right) \tag{7.17}$$

Equation (7.15) gives the field strength in any direction in the plane of the paper due to the two isotropic radiators in terms of the distance between them, the wavelength, and the field strength due to a single radiator radiating the same total power.

In equation (7.16) this is restated as the product of the field strength E due to a single radiator and an array factor. The array factor accounts for the modification of the field of a single radiator due to the radiation of power from the two radiators. If the array factor alone is plotted, what is known as the **'group pattern'** is produced.

Example 7.4

Consider two isotropic radiators spaced by a quarter of a wavelength, as shown in Fig. 7.15, i.e. $d = \lambda/4$. Determine the radiation pattern.

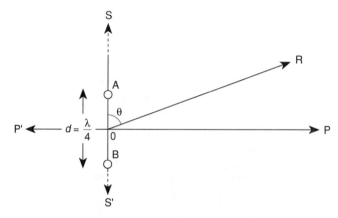

Figure 7.15

Solution
Since $\theta = 90°$, $\cos \theta = 0$

$$\therefore \text{ array factor} = \sqrt{2}\cos\left(\frac{\pi}{\lambda} \times \frac{\lambda}{4} \times 0\right) = \sqrt{2}\cos 0$$

$$\therefore E_\theta = E \times \sqrt{2}$$

When $\theta = 0°$, $\cos \theta = 1$

$$\therefore \text{ array factor} = \sqrt{2}\cos\left(\frac{\pi}{\lambda} \times \frac{\lambda}{4} \times 1\right)$$

$$= \sqrt{2}\cos\left(\frac{\pi}{4}\right) = \sqrt{2}\cos 45° = 1$$

$$\therefore E_\theta = E$$

The intermediate value of E_θ can be found in the same manner. The radiation pattern in the plane of the paper is as shown in Fig. 7.16. This shows that maximum gain is in the directions OP and OP'.

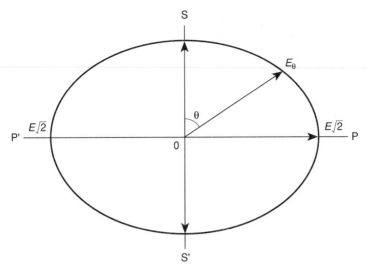

Figure 7.16

Example 7.5

Consider the two radiators when $d = 1$ as shown in Fig. 7.17. When $\theta = 90°$, $\cos \theta = 0$. Determine the radiation pattern.

Solution

$$E_\theta = E\sqrt{2} \quad \text{when } \theta = 90°$$

When $\theta = 0°$, $\cos \theta = 1$

$$\therefore \ \text{array factor} = \sqrt{2} \cos\left(\frac{\pi}{\lambda} \times \frac{\lambda}{2}\right) = \sqrt{2} \cos 90° = 0$$

$$\therefore \ E_\theta = 0 \quad \text{when } \theta = 0°$$

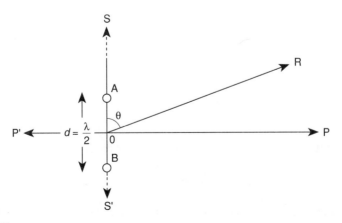

Figure 7.17

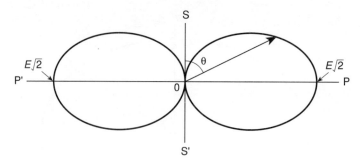

Figure 7.18

The radiation pattern in the plane of the paper is as shown in Fig. 7.18 when the radiators are spaced by half a wavelength. The maximum voltage gain is in the directions OP and OP′ and is equal to $\sqrt{2}$. It is zero in directions OS and OS′. The zero field is to be expected since a wave from B, travelling towards radiator A, arrives at A after travelling half a wavelength and is thus 180° out of phase with the radiation from A and cancels it out. Similarly, a wave from A cancels that from B in the opposite direction.

7.8 Co-linear dipoles

The radiation patterns shown so far are all in the plane of the paper. Considering this plane alone, the value of the field strength in any particular direction is given by the product of the field strength of a single isotropic radiator radiating the same power and the array factor, which accounts for the modification of the radiation pattern due to radiating power from the two elements of the array.

To illustrate how this can be applied to practical antennae a pair of co-linear dipoles and a pair of dipoles placed side by side will be considered. Consider two dipoles in free space spaced $\lambda/2$ apart, fed in phase and mounted above one another as shown in Fig. 7.19(a). The radiation pattern of a single dipole is shown in Fig. 7.19(b). This is in the form of a figure of eight and, to find the value of the field at any angle $\theta°$ to the vertical, the following formula is used:

$$E_\theta = E\left(\frac{\cos((\pi/2)\cos\theta)}{\sin\theta}\right) \tag{7.18}$$

The array factor in the case of two dipoles mounted one above the other is the same as that calculated previously for two isotropic radiators and is given by

$$\sqrt{2}\cos\left(\frac{\pi d}{\lambda}\cos\theta\right)$$

Plotting the array factor gives the group pattern shown in Fig. 7.19(c). The resultant of the two patterns is shown in Fig. 7.19(d) and is found by multiplying

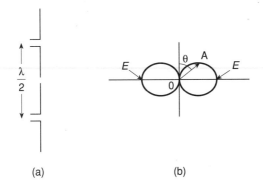

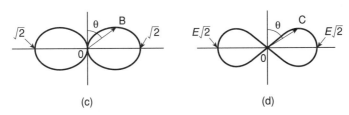

Figure 7.19

the two factors together:

$$E\frac{\cos((\pi/2)\cos\theta)}{\sin\theta}\times\sqrt{2}\cos\left(\frac{\pi d}{\lambda}\cos\theta\right)\qquad(7.19)$$

Referring to Fig. 7.19, OC on Fig. 7.19(d) is equal to the product of OA on Fig. 7.19(b) and OB on Fig. 7.19(c). Looking down on the antenna, the horizontal radiation pattern is a circle of radius $\sqrt{2}E$.

Example 7.6

If two half-wave dipoles are separated by one wavelength, determine the radiation patterns produced.

Solution
Figure 7.20 shows two half-wave dipoles separated by a distance of one wavelength. The single dipole pattern is shown in Fig. 7.20(b) and has the same equation as in the previous example. The group pattern is shown in Fig. 7.20(c). The array factor is as before, but the value $d = \lambda$ is now substituted in the equation. The resultant of the product of the two equations is shown in Fig. 7.20(d). Usually Figs 7.20(b) and (c) are omitted, and Fig. 7.20(d) presented directly as the vertical radiation pattern of two co-linear dipoles fed in phase and spaced one wavelength apart.

   The horizontal radiation pattern would be a circle, as in the previous example of two co-linear dipoles spaced by half a wavelength.

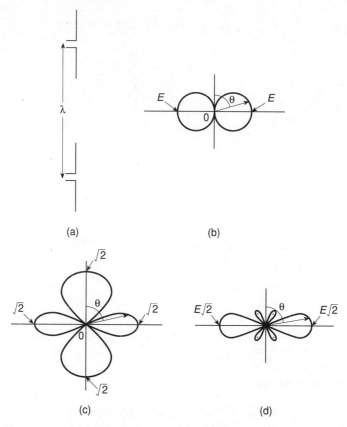

Figure 7.20

Vertical dipoles arranged one above the other are suitable for certain types of broadcasting.

7.9 Parabolic reflectors

Antenna gain has been referred to several times in this chapter and various methods of achieving it have been briefly dealt with. However, antenna gain and directivity can reach high values when using dish antennae. Antenna gain is a measure of its directivity as the main aim is to concentrate most of the transmitted energy in one direction. This is particularly necessary in satellite communications where beamwidths as narrow as 1° are required for accurate reception.

In order to achieve high gain a parabolic dish is used and a dipole placed at its focus. This is very much like concentrating light through a lens or placing a car headlamp bulb at the focal point of a reflector. Practical dish antennae are a little more complex in that waveguides are normally used to feed them.

The parameters of a dish antenna are best understood by considering the geometry of the paraboloidal-shaped dish as shown in Fig. 7.21. In this diagram a

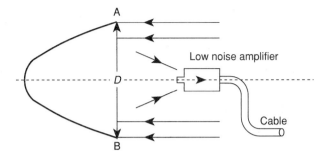

A

Low noise amplifier

D

Cable

B

Figure 7.21

receiving dish is shown, but the principle of operation applies equally well to a large transmission dish. It is found that if the dish is circular instead of parabolic then radio waves parallel to the principal axis will be reflected to different focal points. This is similar to astigmatism caused by an optical lens. Altering the curvature of the lens brings the light rays into focus.

Note that while the dish configuration is in the shape of a parabola, the mouth aperture is circular. The mouth aperture is AB. The aperture diameter (D) should be greater than one wavelength at the operating frequency and the aperture itself is defined by

$$A = \frac{F}{D} \tag{7.20}$$

In practice the F/D ratio varies from 0.25 to 0.5. If this ratio is low, the reflector is not fully illuminated by the radio energy leaving the focal point, i.e. it performs less efficiently. If the ratio is large the reflector loses energy over the rim and this is known as **spillover**. This further reduces efficiency.

As the dish is not fully efficient the aperture is normally referred to as the **effective aperture** (A_e). This is the area which absorbs a power equal to that which is actually received by the dish and hence transferred to the receiver. Hence

$$A_e = \eta A \tag{7.21}$$

where η is the efficiency factor (>1).

The effective aperture and antenna gain are related by the simple but useful expression

$$G = \frac{4\pi A_e}{\lambda^2} \tag{7.22}$$

Also

$$G = \frac{\eta 4\pi}{\lambda^2} \times \left(\frac{\pi D^2}{4}\right) \tag{7.23}$$

$$G = \eta \left(\frac{\pi D}{\lambda}\right)^2 \tag{7.24}$$

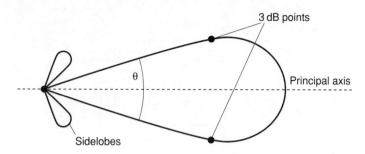

Figure 7.22

The following points should be noted about this gain:

- This is the gain relative to an isotropic radiator. Remember this is a point source radiator.
- It is the gain along the principal axis of the dish, as shown in Fig. 7.22, and hence it is the maximum gain.

If the dish departs from the principal axis then the gain becomes

$$G_{dB} = 10 \log \eta (\pi^2 D^2 / \lambda^2) - 12(\alpha / \theta_{3dB})^2 \tag{7.25}$$

where α is the depointing angle and

$$\theta = 70 \frac{\lambda}{D} = \text{beamwidth at half power} \tag{7.26}$$

Example 7.7

A parabolic dish antenna operates at a frequency of 12 GHz.

(a) Determine the gain of the antenna if it is 65 per cent efficient and has an aperture of 0.785 m.
(b) Calculate the beamwidth.
(c) Calculate the depointing angle if the total power is 19 dB.

Solution

(a) The maximum gain is referred to in this part of the question, hence this will be found as follows with respect to an isotropic radiator:

$$G = \eta \left(\frac{\pi D}{\lambda} \right)^2$$

and

$$A = \frac{\pi D^2}{4}$$

$$D = \sqrt{\frac{4A}{\pi}} = \sqrt{\frac{4 \times 0.785}{3.14}} = 1 \, \text{m}$$

Also since $\lambda = v/f$ then

$$\lambda = \frac{3 \times 10^8}{12 \times 10^9} = 0.025\,\text{m}$$

hence

$$G = 0.65\left(\frac{3.14^2 \times 1^2}{0.025^2}\right) = 102.54 = 20.11\,\text{dB}$$

(b) The beamwidth is given as

$$\theta = \frac{70 \times \lambda}{D} = \frac{70 \times 0.025}{1} = 1.75°$$

(c) The depointing angle can now be found from the results of parts (a) and (b):

$$G = 10\log\eta\left(\frac{\pi D}{\lambda}\right)^2 - 12\left(\frac{\alpha}{\theta_{3\,\text{dB}}}\right)^2$$

$$19 = 20 \times 11 - 12\left(\frac{\alpha}{1.75}\right)^2$$

Transposing gives

$$\alpha = \sqrt{\frac{(20.11 - 19)3.06}{12}} = 0.532°$$

7.10 Polarization

All electromagnetic energy consists of a magnetic field and an electric field, but in electrical field work the electric field is more significant.

Consider a light source such as an electric bulb. In this case the electro-magnetic energy is radiated in all directions as shown in Fig. 7.23(a). In Fig. 7.23(b) the resultant electric field E has been resolved into its vertical and horizontal components. These can be further separated by polarization filters and analysers into separate horizontal or vertical elements. This is called **linear polarization**.

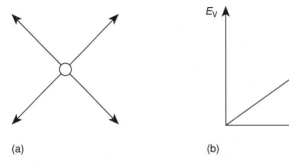

(a)   (b)

Figure 7.23

This principle can be used in multifrequency use for antenna systems. Radio systems use linear polarization and depending on the feed point, they may be horizontally or vertically polarized. This means that an antenna system transmitting and receiving several frequencies at once may radiate these frequencies with different polarization orientations. This gives good isolation of 20–30 dB.

Vertically polarized antennae are generally preferred where long distances are involved and a horizontal antenna system cannot be elevated high enough. Horizontally polarized antennae are preferred for long ranges where high gain is required and low noise transmission is essential.

A gradual shift in polarization in a medium is known as Faraday rotation. The wave rotates 360° as it travels a distance of one wavelength. This is referred to as circular polarization and is frequently used in mobile equipment.

An antenna that is either vertically or horizontally polarized will receive a signal equally well from right- or left-hand circularly polarized transmitting antennae. However, when both the receiving and transmitting antennae are circularly polarized they must have the same sense of rotation.

7.11 Further problems

1. A Yagi array consists of a half-wavelength dipole with a reflector and three directors. Calculate the approximate dimensions and spacings for the elements if the operating frequency range is 640–820 MHz.
   Answer: Reflector length = 2.31 m; director length = 2.09 m; director/dipole = 0.66 m; reflector/dipole = 0.1 m

2. A satellite receiver works in the band 4–6 GHz. If the field strength of the antenna is 1600 μV/m and the voltage induced into the antenna is 1 mV on average, determine the number of elements in the required array.
   Answer: 11 elements.

3. Two dipoles are configured in a co-linear array. Their feed points are separated by a quarter-wavelength. The value of the E field at any angle θ degrees to the vertical is given as

$$E_\theta = \frac{E\cos(\pi\cos +\theta/2)}{\sin\theta}$$

Such an arrangement has to be used for transmitting a frequency of 65 MHz. Determine the value of the field for different values of θ and hence plot the radiation pattern using the appropriate array factor, assuming $E = 800\,\mu V/m$.

4. A co-linear array consists of two dipoles separated by $\lambda/2$. Draw the field pattern for this system if it has to be used for transmitting a frequency of 72 MHz. Hence determine the value of the field when $\theta = 37°$ and $E = 800\,\mu V/m$.

5. A parabolic dish antenna operates at a frequency of 4 GHz.
   (a) Determine the gain of the antenna if it is 55 per cent efficient and has an aperture of 0.8 m.
   (b) Calculate the 3 dB beamwidth.
   (c) Calculate the total power if the depointing angle is 0.12°.
   Answer: 29.9 dB; 5.2°; 29.6 dB

6. A parabolic dish antenna has an effective aperture of $0.65\,m^2$ and operates at 6 GHz. Assuming an antenna efficiency of 55 per cent, determine:
   (a) the diameter of the dish;
   (b) the maximum gain;
   (c) the beamwidth;
   (d) the depointing angle if the resultant gain has to be a minimum of 33 dB.
   Answer: 1.22 m; 35.16 dB; 2.84°; 1.2°

7. A UHF antenna is designed to transmit a signal at 840 MHz. If the antenna current is 2.5 A determine the radiated power of the antenna.
   Answer: 940 W

8. Determine the diameter required for a parabolic reflector antenna working at 2 GHz if its gain is 30 dB. Also calculate the half-power beamwidth.
   Answer: 2 m; 5.2°

9. Determine the lengths of the director and reflector elements used in a 60 MHz Yagi antenna if the folded dipole is a half-wavelength long.
   Answer: Reflector length 2.625 m; distance 1.25 m; director length 2.375; distance 0.75 m.

Index